Embryogeny and Phylogeny of the Human Posture 1

Comparative Anatomy and Posture of Animal and Human Set

coordinated by
Djillali Hadjouis

Volume 3

Embryogeny and Phylogeny of the Human Posture 1

A New Glance at the Future of our Species

Anne Dambricourt Malassé

iSTE

WILEY

First published 2021 in Great Britain and the United States by ISTE Ltd and John Wiley & Sons, Inc.

ISTE Ltd
27-37 St George's Road
London SW19 4EU
UK

www.iste.co.uk

John Wiley & Sons, Inc.
111 River Street
Hoboken, NJ 07030
USA

www.wiley.com

Library of Congress Control Number: 2021938413

British Library Cataloguing-in-Publication Data
A CIP record for this book is available from the British Library
ISBN 978-1-78630-606-7

MIX
Paper from responsible sources
FSC® C013604
FSC www.fsc.org

Contents

Preface

This book is not a new overview of archeological and paleontological discoveries about our origins. It is sufficient to consult the Internet with the right keywords to see the prodigious quantity of articles and books published over the last 10 years on the subject. To assemble so much knowledge would represent an encyclopedic program summing up a history of at least 45 million years, built upon fossilized teeth and bone fragments, karyotypes of current primate species and reconstitutions of locomotor functions, diets, ecosystems, climates, reproductive strategies, subsistence behaviors and, in humans, symbolic behaviors. Furthermore, despite this gathering of information, which is without equivalent in the history of science, such a project would not provide a scientific description of the mechanisms that have prevailed in the emergence of our verticality; not that of the spine, but that of the nervous system, which is seen from the inside of the skull, at the center of the base, and begins with the brain stem, positioning the cerebellum above the pelvis and below the brain. It should be recalled that in a lemur (a current primate very close to the oldest primates), or a rodent, a cow or a dog, the cerebellum is behind the brain and in front of the pelvis in quadrupedal posture.

The human singularity upon which philosophers have focused since Antiquity is twofold: namely, the recognition of this verticality of the nervous system, which does not exist in any other animal, and of the capacity for reflection of the world which has become a reflection on the world and on humanity itself. This book demonstrates that the verticality of the nervous system and the capacity to reason on both sides of reality, the external world and its inner reflection within the human being, are closely linked.

The decisive factor is the position of the cerebellum in relation to the vertical axis of the body, suspended above the spine and therefore very unstable. Its cognitive role in the hominization process was much more important than paleoanthropologists of the 19th and 20th centuries had imagined. As the processes that led to this straightening have no reason to stop, the cerebellum and the neural networks it forms with the brain are logically destined to develop in the future of the physical and psychic postural equilibrium of our species.

This gap on the origins of our vertical posture has been filled over the last 30 years through immersion in some of the largest collections in the world of living primates, including humans and numerous fossils. These collections belong to the *Muséum national d'Histoire naturelle de France* and the *Institut de Paléontologie Humaine* at Paris, a foundation of Prince Albert I of Monaco dating from 1910. This large building, which looks like a palace, is historically the first research center entirely dedicated to the study and understanding of the origins of the human. Its foundation follows in the footsteps of the naturalists of the *Jardin royal des plantes* (Royal Botanical Garden), which became a National Museum during the French Revolution in 1793. These naturalists have sought to understand the place of humans in nature since Georges Buffon (1707–1788), Louis-Jean-Marie Daubenton (1716–1799), Jean-Baptiste Lamarck (1744–1829), Georges Cuvier (1769–1832), Étienne Geoffroy Saint-Hilaire (1772–1844), Armand de Quatrefages (1881–1995), Marcellin Boule (1861–1942) and Jean Piveteau (1899–1991), to name but a few of the main actors in the pantheon of French research. A slow evolution of knowledge combined with critical reflection has allowed a dissociation between the metaphysical conception of the origin of species and the recognition of physical processes in their organization and transformation/adaptation. This "moult" took place at the end of the 19th century, with profound conceptual clashes on the identity of the human within the *Société d'Anthropologie de Paris*, the *Collège de France* and the *Muséum national d'Histoire naturelle*.

The origin of axial verticality is not to be sought in locomotion, which is totally unrelated to the causes of straightening; human bipedalism is one of its consequences among many others, such as the regression of prognathism, the disappearance of the canine tooth (fang) or the shape of the pelvis. This work thus presents a discovery which continues to overturn the mental representation of the origins of our verticality, which unifies the body, the skeleton, the nervous system and the self-consciousness.

Straightening is the consequence of modifications in the development of the embryo along the antero-posterior axis, initially horizontal in all vertebrates, including humans. These modifications began around 45 million years ago in an Asian population of prosimians, the name commonly used to designate the anatomical organization of the oldest primates. One of these species literally gave birth to the first apes, or simians. The difference between the prosimian and the simian is at the highest level of the organization of the species, i.e. its embryogenesis. With the emergence of the simian, the antero-posterior axis of the embryo was broken for the first time in 600 million years, precisely at the center of the base of the skull, so that the cerebellum is slightly lowered forward and downward. With simians, verticality began to become part of the cranio-caudal organization of vertebrates. This is a unique event in their history because what characterizes the vertebrate is the presence of a visible embryonic structure along the embryo, called the dorsal chord. A very small part of the embryo is in front of its cephalic end, which will give rise notably to the two cerebral hemispheres. However, the axis of rotation which allows the straightening is located exactly at the cephalic apex of the dorsal chord. In other words, the straightening process marks a new phase in the evolution of the dorsal chord and its maximum straightening is *Homo sapiens*.

The primitive horizontal organization of the prosimians is still visible in current species such as the lemur (Lemuriformes) and the tarsier (Tarsiiformes). The course of embryogenesis, still common to both of them, has thus been modified during the reproduction of sexual cells, gametogenesis (or meiosis). This elementary deduction would not be surprising if it were an accidental, non-reproducible event. However, it is precisely the opposite. Verticalization was amplified 20 million years later by the birth of semi-erect great apes with a more complex central nervous system. Then, again 16 million years later, the first threshold of verticality appeared within great ape populations. The mutants were the hominids. For the first time in the evolution of vertebrates, the dorsal chord was verticalized. The oldest fossils are known from Africa and date to 4 million years ago, with *Australopithecus* and possibly (and logically) *Homo*. Finally, the last threshold of verticality emerged around 200,000 years ago with *H. sapiens* Linnaeus 1758, or the anatomically modern human.

Straightening is a scientific observation established according to richly documented metric protocols. The fact that it was accompanied by a complexification of the central nervous system, without equivalent in other mammals, is unanimously accepted by paleoneurologists and zoologists.

The reiteration of these processes active during gametogenesis does not correspond to any evolutionary paradigm in force. On the contrary, neither neo-Darwinism nor deterministic chaos theory corresponds to the transmission of mechanisms and their self-memorization. Only the properties of reproduction of reorganization processes can account for the stability of a geometric and dynamic trajectory over such long geological periods.

Scientific induction, which is based on facts and not hypotheses, thus allows us to conclude that gametogenesis also produces its own complexity and self-regulation, including that of molecular networks not limited to chromosomes.

The implications are considerable, as much for the comprehension of the so-called "evolutionary" mechanisms in operation at the time of the gametogenesis which preceded Sapiens as for the recognition of our evolutionary identity here and now, and thus for the future of our posture, which does not separate the body from its conscious reflection (body image, symbolic image) nor from the conscious reflection of the self and its otherness.

The conceptual difficulty of self-memorization begins with the origin of the first cell cycles, or the appearance of life, and this is why, since the philosophers of Antiquity, we regularly find a reflection on the natural links between the origins of life, reflexive consciousness and symbolic thinking expressed in *H. sapiens*.

In a few simple words, the problem posed is the following: how does the process of construction of a closed, dynamic and complex system, such as a cell, attain memory of itself, since its principle is to reproduce? This question is primordial and has been bothering minds since the first Greek philosophers of nature, with the question of the generation of the species.

This book is organized in two volumes. The first follows the progression of the naturalist approach from the pre-Socratic philosophers to the 19th century. The philosophers sought to understand the place of human anatomy in nature, beginning with understanding its laws through measurement, number and equation, which would eventually become the language of science, while the study of anatomy dealt with the animation, movement of organisms and the immutable reproduction of their species. Generation poses the question whose answer is the most elusive: that of the beginning of each individual of the same species that unfolds from the levels, invisible to the naked eye, of the fertilized egg. For all that, these philosophers do not sacrifice reason and do not confuse the number with the psyche that expresses it. The observer of nature is in nature, the psyche is its measure and its reflection, but it is neither its conception (metaphysical reflection) nor its truth (philosophical wisdom).

Understanding the natural order became a necessity with the *History of Animals* of Aristotle, who established the Alexandrian school (Egypt). This naturalist approach disappeared with the collapse of the Roman Empire. The rebirth of the naturalist mind took more than 1,000 years in an intellectual context that became dogmatic with Christian scholasticism. Then, astrophysicists, physicians, philosophers and metaphysicists did not cease fighting to free themselves from it until the 19th century, imposing, *in fine*, the recognition of the natural filiation between the human anatomy and that of apes. However, the majority of so-called "transformist" naturalists did not abandon reflection on the nature of being and of the human being in particular, whether it was Paul Broca in the middle of the 19th century, the founder of biometric anthropology and free thought in Europe, or Albert Gaudry, the designer of the paleontology and comparative anatomy galleries of the *Muséum national d'Histoire naturelle*, at the very beginning of the 20th century. To Charles Darwin's natural selection, Albert Gaudry opposed the "law of love", because if blind nature selects, human beings have a sense of possible otherness that raises the question of values, of their origin as well as their nature, the alterity that humanizes the *H. sapiens* animality.

Understanding the filiation between the individuals of the same species passes necessarily through the study of the generation's processing of each individual. This reflection crosses all the fields of knowledge, confronting the infinitesimal scales of time and space, and passing gradually on to scales visible to the naked eye, following the orderly construction of the organs.

For the naturalists of the Age of Enlightenment (late 17th–18th centuries), the problem of generation was impossible without mathematics, the only tool to describe form at the smallest scales: the point, the straight line, the curvature of the straight line, the circle and the sphere. And of course, the point raises the question of the infinitely small, which refers to the foundations of reality. Is there a physical limit? Atomists have been convinced of this since Antiquity. To form is added the problem of movement, because nothing is static in the physical or measurable world; pure form is mathematical, whereas reality is in motion at all scales. Does movement come first? Descartes thought so, but it was Newton who demonstrated that movement is only one consequence of a physical reality upstream from all: the force of attraction. The physical world is a world of attractions and repulsions, which animate or stabilize. This primary reality has never existed in human eyes otherwise than in the form of an equation.

Let us keep in mind that until the beginning of the 19th century, the cosmos was not a cosmogenesis in progress. The explanations to account for the passage from the infinitely small, present in gametes, to the emergence of an animated and living construction after fertilization by gametes were pure speculation, thought experiments. This problem was most inaccessible to the understanding; even Charles Darwin came up against it with his gemmules theory, following the Lamarckian logic of the heredity of acquired characters.

Everything is a new beginning, everything holds together and everything is movement, attraction and repulsion. Physiology (breathing, blood circulation and vision), the development of the embryo, the fetus and the reproductive organs of women have been approached very gradually since the Renaissance, taking account of physics, which should allow us to understand the rules of functioning of the living being (blood pressure, muscular forces, etc.), and of chemistry, with medicinal plants for better healing.

Until the beginning of the 19th century, the history of animals remained a distribution in time and space, of types, or archetypes, within the limits of which the subspecies could vary but without the passage from one type to another ever being conceivable. The explanation still refers to the ontological essence of the philosophers of Antiquity.

The second volume follows the great conceptual divide generated by the first theory of the simian origins of the human. It was formulated in 1802 at the *Muséum national d'Histoire naturelle* by Jean-Baptiste Lamarck. The divide was not completed before the end of the 19th century.

We will follow the epistemological approach and the development of methods that aim to understand the beginning, no longer of the generation of the same species after fertilization, but that of a new embryonic body plan. This generation of successive plans thus presupposes mechanisms that have united current and fossil species of primates throughout the geological and climatic history of the Earth.

The discipline that focuses on the study of the origins of human anatomy did not exist before the 20th century: this would be human paleontology, a new discipline that benefited from the development of techniques allowing an exponential accumulation of anatomical and metric data, thanks to microscopes, teleradiography and tomography (CT scans) and ever-faster management of data processing by artificial intelligence. Researchers of the 21st century thus have vast resources to continue to understand the place of human anatomy in the natural history of animals, as well observed by Aristotle.

Homo sapiens is not the endpoint of creation as in the biblical tradition, but we are indeed, with the straightening of the embryonic dorsal chord, the last stage of a physical phenomenon that is not observed in any other lineage of mammals. This evolution has followed the increasing complexity of the nervous system, inseparable from its straightening and also from its capacity for reflection and discernment between quantitative physics and qualitative self-awareness. The problem of generation has shifted from the species to that of a process that is self-sustaining and grows like an organism. *Homo sapiens* is thus all the more concerned since this pertains to its natural identity and its links with other animal species, whose past traces remain in the gametes. Having reached the summit of verticality – blocked by the face that has contracted in the opposite direction to the straightening – it is now the future of this straightened nervous system that will be discussed, and therefore the future of the reflected consciousness, both individual (symbolic image of the self) and collective (symbolic image of otherness), and which is constructed during psychomotor development. This intuition of a growing super-organism was that of Leibniz in the 17th century and of Teilhard de Chardin in the 20th century. It has lost none of its relevance and continues to challenge us, sending back to us an awareness that gives meaning to the notions of irreversibility, alterity and responsibility.

May 2021

PART 1

The Vertical Human: Philosopher of Nature

1

Anthropos, the First of the Animals

1.1. Introduction

By *anthropos*, we mean the living human being at every stage of its development since fertilization without distinction of gender.

Vertebrate paleontology and comparative anatomy have been elevated to the rank of natural sciences thanks to the collections of the *Muséum national d'Histoire naturelle* and, above all, the vast program of Georges Cuvier (1769–1832) which aimed to extract from them a systematic understanding of life in the past and present. His courses at the *Collège de France* between 1829 and 1832 were dedicated to the history of natural sciences since the pre-Socratic philosophers, or more precisely, to the history of the human psyche confronted with the need to give meaning, and thus to put in order, everything accessible to the senses. His lectures were carefully noted by an assiduous listener, T. Magdeleine de Saint Agy. The *Muséum national d'Histoire naturelle* published the first English translation, preserving the French texts, annotated by Théodore Pietsch (2012, 2015).

This sum of knowledge is the first epistemological study of natural history and the most complete that can be conceived for its time; it was yesterday in the scale of history, the beginning of the 19th century.

The sense of observation manifests the need for every living organism to remain informed of the natural laws that keep it alive or pose a threat. Human beings are no exception, but they possess the ability to understand these laws, either to protect themselves from them or to try to appropriate and control them, endowed with capacities of anticipation unequalled in the animal kingdom. This sense of observation is the descriptive study of the thing being questioned right down to its organization in order to infer principles or laws, whether it be the stars, the seasons, geological strata, a mineral, a plant or an animal, the seasonal course of a river or earthquakes. Since Antiquity, these observers-scrutinizers of nature have been aware that they were the only organisms driven by this need to understand and possess the keys to knowledge; they were its measure, gifted with the means to control, predict and heal.

Thus, the human being has recognized itself as the reference in the world by which the world may be measured and understood. Cuvier sought to highlight the circumstances that have fostered, or conversely hindered, the natural desire of human beings to understand and signify their own nature as well as nature as a whole. His investigation stretched as far back in time as possible, as far back as Antiquity, following a critical approach to the environmental and societal context. The geological reference is Diluvium, the penultimate age of the Earth, characterized by floods on a planetary scale covering the great river plains with their fauna and flora, such as the Indus and Ganges basins. This geological period corresponds to the very ancient Quaternary period that preceded the great rise in sea levels, a geological period that would therefore have preceded the appearance of the human according to Cuvier's contemporary naturalists. Today these are the geological periods of the Pleistocene, which began around 2.58 million years (Ma) ago, followed by the Holocene, beginning with the end of the Ice Age 10,000 years before the present. The post-glacial period was therefore a period of global warming that caused the rapid rise of waters.

Georges Cuvier's epistemological synthesis is an unparalleled break with the centuries that preceded it. It is also a major reference for resituating the conceptual and intellectual contexts in which Jean-Baptiste Lamarck ventured to formulate the first mental representation of the transformation of primates into *H. sapiens*. The French naturalist focused on exclusively natural causes, i.e. physical causes, dependent on long geological periods, climate change and the natural causes of movement.

Referring to Georges Cuvier helps us to also understand the scientific and conceptual difficulties which have been posed by transformism as it has been applied to the origins of humanity for more than two centuries, and which are still unresolved. In fact, there was never any question of the segmented axial endoskeleton, i.e. the bone tissues that surround the nervous system located under the cerebellar tentorium, namely, the cerebellum, the brain stem and the spinal cord. These are the territories concerned by the dorsal chord above and on either side of which the skeleton forms segments. These areas do not begin with the first cervical vertebra, but with a bone in the center of the base of the skull, the basi-sphenoid, flanked by the petrous pyramids.

Thus, the great contemporary epic narratives never show the inside of the axial skeleton, the innermost space of the body, whereas it is from the inside that everything begins with the embryo. These narratives concentrate on external appearance, giving it the driving role in transformation, ignoring the presence of a particularly complex internal organization that follows its own embryonic and phylogenetic constraints, that is to say, those transmitted through heredity and which also follow their own rules.

It is therefore important to follow the progression of this fundamental research since Georges Cuvier's synthesis. He did not support Lamarck's overly simplistic thought experiment, of the quadrupedal adult monkey which became upright. However, due to lack of knowledge in molecular biology and biophysics, he opposed the possibility of natural evolution by making the choice of the metaphysicists of Antiquity, namely, that the human is an embodied "idea", just like the great apes.

1.1.1. *Epistemology according to Georges Cuvier*

For Cuvier, the natural sciences appeared with the oldest societies of pastoralists and farmers established in the flood plains of northern India, those of the Ganges and then the Indus. These societies survived Diluvium by taking refuge in the Himalayan and Tibetan highlands. The river plains of Mesopotamia and the Nile would have been populated only after this geological period. Cuvier thus considered India to be "the birthplace of the human species and of civilization" (Cuvier (1841–1845), Lesson 4 on the Greeks). The naturalist noted, however, that no progress in the natural sciences could be observed since this high period of Hindu Antiquity, and he attributed responsibility for this to the priestly and dominant class of the Brahmins. He interpreted this resilience as the consequence of the religious postulate that the temporal is of no importance: "the age of earthly life is said to be a despicable age unworthy of being preserved in the memory of men". The world is corrupt, the temporal does not deserve the interest of people; history does not exist and the body has no earthly history. The class of intellectual elites therefore did not develop a natural history, that is to say, a reasoned catalog of the plant and animal kingdoms. Knowledge in chemistry was very limited and did not concern the body because of the religious taboo forbidding touching corpses, those of humans being consigned to cremation. Cuvier had knowledge of the upavedas written in Sanskrit and which included knowledge relating to science (astrophysics, methods of calculation), medicine, architecture and mechanics. These texts are then dated to about 20 centuries BCE (Before Common Era) and are the oldest written trace of a scientific knowledge frozen for more than four millennia:

> These obstacles to the development of science still exist today in all their energy; for lately a Brahmin has only agreed to give a book of the Vedas to an Englishman on the express condition that he would not have it bound in moroccan or leather. With such prejudices it is

> impossible to form zoological collections, and science is stopped for ever. (Cuvier, *Histoire des sciences naturelles*, 1841–1845)

Cuvier continued his investigations and deduced that these Brahmin Indians civilized the populations of Egypt by transmitting to them an organization of society in a class system. On the other hand, the Egyptians had a pantheon of animal-headed deities inherited from practices observed among African hunting peoples. These traditions would have promoted the development of comparative anatomy through the mastery of embalming human bodies and a large number of animal species (birds, reptiles, carnivorous mammals, herbivores and monkeys), both divinized and profane. This religious freedom allowed the blossoming of the natural sciences, parallel to the spatial sciences with geometry for the measurement of fields, to the sciences of chemistry for colors, medicine and embalming of bodies. To conclude: "the origin of natural history and anatomy owes most to Egypt".

Cuvier continued his investigations with his reading keys. Natural history could not develop until a new intellectual posture was observed among the philosophers of nature. It was in the Greek colonies that schools of thought emerged, questioning the world and nature, seeking to approach their foundations or essence. The sciences were introduced there following the sojourns of individuals in the schools of Egyptian priests, and also from Phoenicia, Chaldea and trips to the Caucasus, Northern Europe and Asia.

These Greek schools developed because of two favorable factors, the first being the absence of dominant priestly classes, jealous of their religious and scientific knowledge, which constituted an obstacle to the development of knowledge. The second was the rugged topography of Greece with its mountains and valleys that were not easily accessible, which provided a natural bulwark against invasions, reducing periods of unrest unfavorable to the development of the mind's pre-occupations. In fact, the first schools appeared in its colonies along the Mediterranean coast in Asia Minor and in southern Italy and Sicily (or Magna Graecia) (Figure 1.1). Continuing his reasoning, Cuvier inferred that the great river plains in India, Babylonia and Egypt were exposed to ravages, looting and destruction of the libraries that preserved sacred texts and scientific knowledge. The Greek cities, starting with those of the colonies of ancient Greece, thus allowed the development of sciences because they were not under the control of a religious power more attached to myths than to empirical knowledge, i.e. based on observation described according to a method that guarantees objectivity. The method can be based on experiment to the extent that irreversibility is not an obstacle. Cuvier recalls four pre-Socratic periods, the oldest of which dates back to 1500 BCE. They gave birth to the most flourishing school of philosophers of nature, the Athens school, from Socrates to Aristotle.

For the philosopher, *anthropos* was the center of the reflection of reality. It recognized itself as the only living being capable of observing nature in order to extract rules, principles and infer knowledge accessible to abstraction capacities, and this in order to derive concepts, principles or main ideas. Different philosophical schools of thought have debated this possibility for the human mind, questioning whether it is possible to access the understanding of nature and the diversity of plants and animals by the mind's intellectual capacities alone or, on the contrary, whether it is necessary to define the outline of its limits by admitting the existence of a beyond inaccessible to the five senses, that of the world of ideas accessible only to the being (meta-physis) incarnated in nature; this state of affairs is called "*physis*" (or "*phusis*"). This term means "beginning" or "generation"; it is the question of the generation of the species. Where does it come from and how does it begin? It is thus that science studies the "how", or physics, while reason continues its natural inclination to search for meaning, the "*raison d'être*", or the finality of necessity, the "why".

Georges Cuvier reminds us of these differences with their particularities.

1.1.2. *From the metaphysics of beings to the physics of their matter*

For contemporary anthropology, it may seem surprising that the questioning of certain philosophers of the human ability to understand nature and its foundations has not been accompanied by a reflection on the physiology of the sense organs and the functioning of the brain. However, naturalists considered it to be the organ of thought as early as the 6th century BCE with the physician-philosopher Alcmaeon of Croton, a Greek colony in southern Italy. These abilities were speculative and often sexist (Sartori 2006).

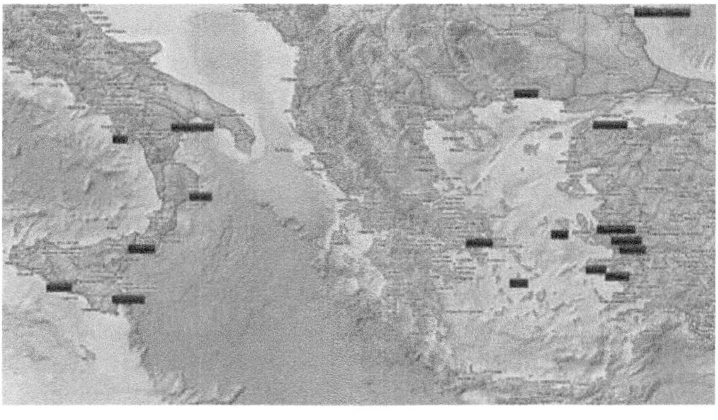

Figure 1.1. *Map of pre-Socratic schools (credit: Kiro Vermaas 2017). This work is licensed under the Creative Common Attribution-Share -Alike 4.0 International license*

The first school was established in Miletus on the coast of Asia Minor; the Milesian school was that of the famous mathematician Thales, born 600 BCE, joined by Anaximander and also Anaximenes, who would have invented the word “world”. They were the first physicists, or naturalists, those who sought to understand the processes that generate movement and organization. Nature was considered perpetually recommencing, while species on the other hand reproduced identically from generation to generation.

Their preoccupations consisted of understanding what animates or is animated, what organizes or is organized, what diversifies or is diversified, without reference to narratives such as myths constructed by the imagination. These philosophers postulated that nature must be understood for what it is based on the five senses and human reasoning; the body and psyche are therefore consubstantial with what is perceptible. Their program was to verify this not only through observation (astronomy, meteorology, etc.) by relying on their senses, but also through demonstration, that is to say, through geometric reasoning constructed with numbers. Thus, objectivity can be considered to have been introduced through mathematics, which was no longer simply to count goods and measure land, but to understand reality for what it is.

According to the oral tradition of the ancient Greeks, or their legends, Thales went to Egypt and measured the height of the pyramid of Giza at the request of Pharaoh Amasis. Verticality was considered the third dimension of space and the most inaccessible, especially in Asia Minor where the snowy peaks of the Anti-Taurus were among the highest (often 3,000 meters and up to 3,700 meters). It was the third dimension that enabled going from surface (the desert, the sea) to volume. Thales had the idea of measuring the projection of the shadow of the pyramid on the ground by taking as a unit that of its shadow in vertical position. He would have found that, whatever the time of day, the projected shadow is proportional to the height of the immobile object. He would then have been able to measure the length of the shadow of the pyramid from its base to its top, after having measured his own shadow, and to apply the ratio of proportion to deduce the height. These were the first equations or equivalences between numbers.

Whatever the degree of veracity or symbolism in the human as the measure of reality by the number of its shadow, it is the problem posed and its solution that matters.

This is the issue that encompassed the scientific understanding of the world, including the sky. This necessarily involved the study of light, because it is light that shows nature, only during the day (or by full moon), and it also involves astronomy, which shows movement in the sky by posing two problems: first, that of the origin of the movement and, second, that of identifying a reference point outside the movement to measure it.

The length of the shadow varies with the relative position of the Sun and the Earth, and also according to the latitude. As these seafaring people who navigated the Mediterranean Sea found, the dimension of the shadow changes and the Earth is not a flat surface, it is curved. The line, the surface, the third dimension and the curve are already four pieces of physical data constituting a key question: what is the curve, why is a line curved? The measured shadow is a line, but later the question was asked differently: is it a line or a curved segment of a sphere since the Earth is spherical? Was this segment static since it moved in space with the rotation of the Earth on its axis and around the axis of the Sun? These questions were asked periodically over the centuries, with Kepler, Copernicus and Galileo in the 16th century, Descartes then Newton in the 17th century and Einstein in the 20th century (the curvature of space-time under the effect of gravity). In the 21st century, it is the notion of space-time and duration that is revisited at the smallest scales when gravitation disappears (Aveline *et al.* 2020).

The nature of matter was therefore necessarily speculative for these philosophers. The difficulty was to understand the cause of movement. Raw matter was necessarily motor and movement, a permanent recommencement. For Thales, this matrix was water, said to have produced everything that is observable. For Anaximenes, solid or hard materials were derived from air. This idea was inspired by observing the formation of clouds suspended in the air, followed by their transformation into rain which solidified as snow, hail and ice on the summits of Mount Taurus. Air could therefore become rocky or earthy and could also become fire by expansion.

Anaximander noted the insufficiency of the four elements, air, earth, water and fire, in accounting for the permanence of reality, since water extinguishes fire, while fire makes air scarce and evaporates liquid water. He inferred from this a level of reality at the origin of everything, which he called *apeiron* (ἄπειρον). This concept was no longer that of the physicist who geometrized nature with its shapes, edges and limits, in plants and animals alike. This approach was no longer that of observing nature through the senses or of demonstration through geometry and calculation, but an abstraction to give a coherent intellectual foundation to what is beyond the observable and the measure.

Doxographers[1] showed the difficulty of classifying this term *apeiron* in either of two categories, ontology or metaphysics (root causes). Anaximander's posture consisted of asking "how something is possible"; it was not underpinned by the search for a meaning referring to a finality. It was a metaphysical posture, seeking

1 Historians of science dating back to the end of ancient Greece, who studied and compiled the philosophers of Antiquity.

explanation, not meaning, whereas the philosopher sought being, the source of meaning.

Georges Cuvier noted well that the notion of creation, proper to the biblical tradition, was absent in this tradition. Moreover, none of these philosophers distinguished between spirit and matter. Jean-Pierre Vernant (2018) writes:

> The work of the Milesians represents a radical innovation: neither singers nor poets nor storytellers, they express themselves in prose, in written texts, which do not aim to unroll, in the line of tradition, the thread of a narrative, but to expose, concerning certain natural phenomena and the organization of the cosmos, an explanatory theory. From the oral to the written, from poetic song to prose, from narration to explanation, the change of register responds to an entirely new type of inquiry: new in the object it designates: nature, phusis; new in the form of thought that manifests itself in it and which is all positive (…). The Milesians do not involve any supernatural being in their explanatory schemes. With them, nature in its positivity has invaded the whole field of reality; nothing exists, nothing has happened nor will ever happen that does not find in the phusis, as we may observe it every day, its foundation and its reason. It is the force of the phusis, in its permanence and in the diversity of its manifestations, that takes the place of the ancient gods. (Vernant, "Les origines de la philosophie", 2018, author's translation)

The physical science that deals with humans is medicine. The first Greek school was founded in Knidos in 700 BCE, on the Mediterranean coast of Asia Minor, where one learned mainly how to heal diseases and war wounds. A new school was established 200 years later in Cos with Hippocrates (460–370 BCE).

1.1.3. *Mathematics, forms and women physicians*

Cuvier continued his course. The second school of the philosophy of nature was founded by Pythagoras (550 BCE); it was faithful to the secret traditions of the Egyptians and was interested in arithmetic, geometry and medicine. Arithmetic calculates (the number of sacks of wheat) and geometry measures (the length or area of a field). Exiled in Magna Graecia, this school of thought considered number as the first principle. The Greek mathematician insisted on the discontinuous aspect of nature; reality breaks down into discrete geometric elements (point, line and surface), which reveal resonances between the harmony of numbers and substance. An observation is established when geometrical constraints are imposed upon proportions in matter, with the idea of harmony; it becomes musical with the strings of a trigonon (a harp) which differ in their proportion and maintain sonorities accepted by human nature. Cuvier supposed that Pythagoras invented mathematical physics in order to be able to compare the physical properties of materials by means of number. Nature would therefore be structured by laws uniting mathematics and the physical world. This school was a religious and scientific sect practicing the culture of secrecy reserved for initiates. This school was also interested in human anatomy.

Alcmaeon was a pupil of Pythagoras (de Wit 1992) and was the first known author of a treatise on human anatomy written with the aid of dissections. He extended anatomical knowledge to the sensory organs, describing the auditory canal and the eardrum. He recognized the brain as the organ that centralizes sensations and perceptions with hearing, smell, taste and sight, and which is therefore the center of common sense and thought. Historians report some remarks on reproduction and embryology; the embryo is said to be formed from both male and female semen, head first. The embryo feeds in the mother's body like an absorbent body, like a sponge (already exploited by Greek sailors for its absorbent properties).

The school of Pythagoras was open to women and defended a conception of education and morals aimed at a certain asceticism:

> Faced with the speculations of the Milesian School, which were essentially about physics, Pythagoras is concerned with Man (…). He drew much of his doctrine from the teachings of Themistoclea, a priestess of Delphi. (Jufresa, "Savoir féminin et sectes pythagoriciennes", 1995, author's translation)

Pythagoras' wife, Theano, was a philosopher, mathematician and doctor, and she took charge of her husband's school after his death. Many women were teachers or scholars and defended the rules of the community. According to the list of Iamblichus (between the 3rd and the 4th centuries CE (Common Era)), the sect gathered 218 males and 17 females, names that influenced their era. John Estobaeus (5th century CE) added four more with Perictione, Melissa, Phyntis and Aesara. Monserrat Jufresa (1995) reminds us that, of their contributions:

> The most important fragment is from the book *On Human Nature* by Aesara of Lucania. It deals with the nature of the soul. Aesara considers that, through a thorough exercise of introspection, we can discover the law and justice that reside there and which offer us the model that should be applied to the home and the city. These virtues, which belong specifically to Man and not to other living beings, stem from the relationship that is established between the parts of which the soul is composed: law and justice are the product of the harmony that

> is established between them. (Jufresa, "Savoir féminin et sectes pythagoriciennes", 1995, author's translation)

The third school, called "Eleatic", was contemporary to Pythagoras. It is said to have been founded by Xenophanes, born 570 BCE in Colophon in Ionia (Asia Minor) but who settled in the south of Italy in Elea (Eleaea). For Cuvier, his doctrine was influenced by the Indian tradition according to which the senses do not reflect reality but are a source of illusions. According to doxographers, the founder was Parmenides, a native of Elea. This school is in reality only known from fragments of a long poem "On Nature" which has given rise to many interpretations that are still relevant today. The notions of being and world are explained for the first time in the history of Greek philosophy. Being is one, indivisible, without beginning or end; it does not last but is limited. The world-being is like a sphere and time is foreign to its nature. The school of physicists was confronted with the difficulty of reconciling being and change, becoming, which is characteristic of the physical world in movement, with animate organisms – animals – that reproduce themselves in their own likeness, whose individuals are sick and mortal, with desires, impulses and feelings.

Rejecting observation as immediate access to ultimate truth, this exercise of reason could only test the solidity of its speculations by imposing upon itself a critical analysis built on logic: a truth cannot generate contradiction. This is the elaboration of a critique of knowledge. Reason therefore imposes the principle of non-contradiction on itself and adopts dialectic as its modus operandi. This exercise of logical reasoning must therefore allow the distinction between a statement that is true in itself and an opinion exhausted by contradiction.

This is notably the case between the discontinuous reality of Pythagoras and the sensory or intuitive experience of the continuous. Parmenides was not a Pythagorean; for him, it was not number that was primary but being. He suggested that discontinuity was only an illusion due to the limited capacities of human reason to apprehend everything, whereas the access to true, immutable, eternal being would be the union of the soul (the sensitive part of the form) with being, it would be continuity in essence. Only reason would be able to perceive it beyond physical perception. The five senses would not give access to the essence of the realities but only their appearance, while the soul is essential and united with being.

Zeno of Elea succeeded Parmenides. The observer being external to the observed object, doubt remained as to the correctness of perception. For lack of being able to decide between the true and the false, or truth and opinion, Zeno offered the possibility of preserving doubt: uncertainty. This posture was that of skepticism.

Melissus of Samos was the last of the Eleatics with his essay "On Nature or What Exists". He returned to Parmenides' contradictions by postulating that being is not like an immutable sphere but, on the contrary, an infinity both spatial and temporal. Infinity is opposed to the finite sphere, so that being unfolds in an infinity of manifestations. Cuvier concluded in his Lesson 5:

> The Ionian school was entirely materialistic, it had not discovered any regulating intelligence: the first Pythagoreans rose above it, seeking and indicating a cause superior to matter. (Cuvier, *Histoire des sciences naturelles*, 1841–1845, author's translation)

The fourth school, known as "atomistic", was quite different. This was illustrated by Leucippe and Democritus (de Buzzon 1999). These two philosophers, active around 440 BCE and 433 BCE, broke with the Eleatics and postulated that reality was constituted of finite physical structures called atoms, distinct forms, and which were unbreakable, separated by the vacuum, in order to be able to justify movement. Invisible to the naked eye, they constituted the smallest level of organization in nature. For Cuvier, the rationality of this mode of reasoning paved the way for modern science.

1.1.4. *Socrates, Plato and Aristotle: an anthropology of ideas or a certain idea of anthropology*

The four schools were distinct and crossed their natural or metaphysical philosophies. Socrates summarized them and passed them on to Plato. Plato's follower, Aristotle, dissented.

Plato postulated the primacy of the ideas that regulate the world of the living and the dead and was content with this; nature was not observed for itself. His anthropology, on the other hand, was sectarian; women, slaves, children and the lower classes of society had only primary needs that were reduced to "desires, pleasures and regrets". Aristotle considered the Platonic universe as a speculation of the psyche on nature before having even studied it, before beginning to be concerned about it or interested in nature for what it is. He did not reject metaphysics, but he considered this mode of knowledge to be arbitrary. He chose observation to understand the ordering of the world, its climatic, seasonal and geophysical diversity, with mountains, fresh or marine waters, air and subterranean worlds, as well as the organization and functioning of living beings that reproduce and flourish in these varied environments. Georges Cuvier was full of praise:

> Aristotle, that vast and prodigious genius, distinguished very well the intertwined branches of science; he classified them methodically with

> an admirable superiority of view, and gave excellent rules for their study. Many of them were also very strongly extended by his discoveries; and he would have given his name to the third epoch in the history of science, if he had had successors able to follow him in his rapid progress. But, by a fatality that happens all too often when a man is much superior to his century, Aristotle had no disciples capable of completing the scientific monument he had begun. The sect he founded, which was called the Peripatetic school, even fell into contempt, and it was not until the 16th century CE that his method was employed, that is to say, that men devoted themselves especially to pure mathematics, astronomy, mechanics, chemistry, physics, etc., and that these sciences, in spite of false directions, made fairly rapid progress. (Cuvier, *Histoire des sciences naturelles*, 1841–1845, author's translation)

This approach is at the origin of the first ordering of the animals according to anatomical or physiological criteria. The work written in Greek around 343 BCE is entitled *History of Animals* and consists of nine books divided into chapters. The human is found among the viviparous and furry animals (animals with udders), but it is also put with the birds according to the locomotor criterion, as both have feet. However, Aristotle needed a reference point to establish an objective organization and this reference point is Man, as he specified in chapters VI and XII of Book 1:

> To begin with, we must take into consideration the parts of Man. For, just as each nation is wont to reckon by that monetary standard with which it is most familiar, so must we do in other matters. And, of course, man is the animal with which we are all of us the most familiar. Now the parts are obvious enough to physical perception. However, with the view of observing due order and sequence and of combining rational notions with physical perception, we shall proceed to enumerate the parts: firstly, the organic, and afterwards the simple or non-composite. (Aristotle, *History of Animals*, 343 BCE, D'Arcy Thompson translation, 1910a)

Henri Daudin reminds us of this:

> Aristotle deems it indispensable, even natural, to begin with Man in the enumeration of the "parts" of the body, as he presents himself to be considered from the outside, and he indicates the reasons for this: it is that we can only appreciate the meaning of the least known from the most known, and that man is necessarily, of the animals, the one that is most known to us. (Daudin, *De Linné à Jussieu : Méthodes de la*

> *classification et idée de série en botanique et en zoologie (1740–1790)*, 1926, author's translation)

Anthropos was the observer and therefore the reference for measuring differences. It is positioned first, followed in parallel by the species that most resemble it, the monkeys. The name "primate" did not appear until 1793 with Joseph-François van der Stegen de Putte (1754–1799); it simply signifies that these mammals have been first in the order of the scale of comparison of animals since Aristotle.

Similarly, according to Balan (1979), who traced the evolution of concepts of order and time from Antiquity to the 19th century, Aristotle's classification was not serial, it was not a scale from the most complex to the least complex, but had the form of categories organized in book chapters that followed one another.

From the point of view of the metaphysics taught by the Socratic school of Athens, Aristotle kept the logic of the philosopher of nature; he sought its principles and essences. Nature was interpreted as the projection of a metaphysical world composed of ideas. Nature was the actualization of the being that exists in itself, for itself and that maintains itself in this kind of harmony closed in on itself. Species did not have a long history and the notion of finality was absent. In such a context, the anthropocentrism criticized at the level of animals as being imbued with a metaphysical finalism was all the more unfounded since finality was a concept foreign to Greek philosophy, which did not conceive of a beginning and therefore of a divine project. Aristotle's hierarchy concerned the souls contained in the semen, supposed to animate the morphogenesis of the species that it carries to its adult stage. Each soul is the bearer of a finality or a plan that is the physical formation of a species until its end. However, these finalities are closed in on themselves, they are not part of a cosmic finality.

Still, according to Balan, it was the Hebrew and Hellenizing philosopher Philo of Alexandria (20 BCE–45 CE) who established the concordance between Aristotle's catalog and the Genesis of the biblical tradition, taking up again Jacob's ladder of the angels, which symbolized the intermediate world between the Earth and the Heaven. His anthropology, on the other hand, was not as good as that of Pythagoras since it was ontologically sexual. Man was the only one to transmit the "soul" of the species; Man was therefore the father of humanity and, by projection, only a father could have created the World. Patriarchy dominated as in Aristotle's work.

1.2. *Anthropos*, the axis of the world

> Of all the sciences, the one that owes most to Aristotle is the natural history of animals, which arranges facts, not according to species, but according to organs and functions, which is the only way to establish comparative results. (Cuvier, *Biographie universelle*, 1811–1828, author's translation)

1.2.1. *Man, a vertical anatomy*

In Chapter XII of Book 1, Aristotle makes an inventory of the external and internal characters of Man that distinguish him from other mammals. The first remark is the vertical organization from the top of the skull to the two feet, "in relation to the axis of the world" (Barthélemy-Saint-Hilaire 1883):

> In man, above all other animals, the terms "upper" and "lower" are used in harmony with their natural positions; for in him, upper and lower have the same meaning as when they are applied to the universe as a whole. In like manner the terms, "in front", "behind", "right" and "left" are used in accordance with their natural sense. But in regard to other animals, in some cases these distinctions do not exist, and in others they do so, but in a vague way. For instance, the head with all animals is up and above in respect to their bodies; but man alone, as has been said, has, in maturity, this part uppermost in respect to the material universe. (Aristotle, *History of Animals*, Chapter XII, Book 1, 343 BCE, D'Arcy Thompson translation, 1910a)

For Aristotle, the axis of the world was the axis of the Earth considered as a static sphere as well as the center of the cosmos around which the stars and planets revolved. This axis was the vertical which also corresponded to the trajectory of the fall of heavy bodies, the axis of gravity due to gravitation, as already conceptualized. The anatomical axis of *anthropos* coincided with the axis of the world and the axis of the Earth's gravity; it was heavily symbolic because the orientation of the brain toward the supra-lunar cosmos polarized it toward the sphere of divine perfection:

> The parts, then, that are externally visible are arranged in the way above stated, and as a rule have their special designations, and from use and wont are known familiarly to all; but this is not the case with the inner parts. For the fact is that the inner parts of man are to a very great extent unknown, and the consequence is that we must have recourse to an examination of the inner parts of other animals whose nature in any way resembles that of man. (ibid.)

Aristotle followed the order of verticality, from top to bottom:

> In the first place then, the brain lies in the front part of the head. And this holds alike with all animals possessed of a brain; and all blooded animals are possessed thereof, and, by the way, molluscs as well[2]. But, taking size for size of animal, the largest brain, and the moistest, is that of Man. Two membranes enclose it: the stronger one near the bone of the skull; the inner one, round the brain itself, is finer. The brain in all cases is bilateral. Behind this, right at the back, comes what is termed the "cerebellum", differing in form from the brain as we may both feel and see. (ibid.)

The naturalist philosopher located the center of intelligence in the heart, but his natural classification already led to him seeing the brain as the organ directed toward the divine spheres, its study considered to be the noblest of the human sciences. The cerebellum, placed differently, in an inferior position, pushed down toward the Earth, had a less noble conceptual destiny until the end of the 20th century.

1.2.2. *Apes and humans*

The zoologist discussed apes in Chapter V of Book 2. The Greek name to describe these hairy quadrupeds is that of an animal called *pithekos* – πίθηκος – which Aristotle generalized to two other animals. He considered them as intermediaries between Man and the other quadrupeds, qualifying these characters as anthropoids (ἀνθρωποειδεῖς). This was the first anthropomorphic formulation of apes, or *simia*, which is the Latin translation of *pithekos*. The term anthropoid was still used, but it no longer referred to those popular monkeys described by Aristotle, macaques and baboons:

> Animals intermediate between man and the quadrupeds; the monkey, of three species; its relationship with the human form; its bestiality; particular organization of its feet, which are feet and hands all together; it walks much more often on all fours than straight. Some animals have a nature that is at the same time that of Man and that of the quadrupeds; they are the *pithekos* (πίθηκος), the *kebos* (κῆβος πίθηκος) and the *cynocephalos* (κυνοκέφαλος). The *kebos* is a *pithekos* with a tail. *Cynocephali* have the same shape as the *pithekos*, except that they are larger and stronger, and their face looks more like a dog's. Their character is wilder; and their teeth, which are closer to the dog's teeth, are also stronger. The *pithekos* is hairy in the upper

2 Aristotle referred to the nervous system formed of ganglia in molluscs.

> parts, because it is a quadruped; and the lower parts are hairy too, because it resembles Man. As has been said above, *things are in Man just the opposite of what they are in animals*. Their face has many analogies with the human face; their nose, ears and teeth, are very similar to those of Man, both for the front teeth and the molars. While the rest of the quadrupeds do not have eyelashes on both eyelids, the monkey has them; but these eyelashes are very rare, especially those at the bottom, and excessively short. The other quadrupeds have none at all. (Aristotle, *History of Animals*, Chapter V, Book 2, 343 BCE, translation of the French version)

The *pithekos* and the *cynocephalos* were tailless monkeys, while the *kebos* was a monkey endowed with a tail. They were the small apes of Africa and Eurasia. The species of the circum-Mediterranean countries known at the time were the *cynocephalos* (*Papio cynocephalus*) and tailless baboons (*Papio hamadryas*), as well as the *cercopithecus* with a tail (close to the macaque). The Egyptians mummified *cynocephalos*, which was divinized into the god Thoth, and sometimes the *cercopithecus* (Lortet 1905):

> The ape has also hands and fingers and nails like man. Its feet are exceptional in kind. That is, they are like large hands, and the toes are like fingers, with the middle one the longest of all, and the under part of the foot is like a hand except for its length, and stretches out towards the extremities like the palm of the hand; and this palm at the after end is unusually hard, and in a clumsy obscure kind of way resembles a heel. The creature uses its feet either as hands or feet, and doubles them up as one doubles a fist. Its upper-arm and thigh are short in proportion to the forearm and the shin. (Aristotle, *History of Animals*, Chapter V, Book 2, 343 BCE, D'Arcy Thompson translation, 1910a)

Apart from the head, the common features shared with humans upon which Aristotle insisted were the anatomy and manual morphology of the feet of these quadrupeds. The great apes of the tropical forests, i.e. chimpanzees and gorillas in sub-Saharan Africa, and the orangutans of the islands of Southeast Asia (Java and Borneo) were not known. For Jules Barthélémy-Saint-Hilaire, there was no doubt that Aristotle dissected monkeys given the accuracy of the descriptions. However, the books preserved over the centuries do not testify to a search for the meaning of these four viviparous quadrupeds with a face and palms. The philosopher may have compared the internal anatomy of the skulls of humans and monkeys – the opposite would be surprising – but no trace of it remains. We would have liked to know his opinion on the coincidence between the vertical axis around which the anatomy and growth of Man is organized and the axis of the world. The comparison of vertically sawn skulls would have shown the differences in the orientation of the cerebellum and the brain stem that extends the spinal cord. Perhaps he would have concluded that the vertical organization of Man can only be understood from inside the axial skeleton. Perhaps he would have formulated the regret of not being able to look at this form in Man from the embryo. Would he have imagined it verticalized from its first physical manifestations?

Aristotle was interested in the comparative anatomy of the stages of growth and he saw that development brings animals closer together as observation approaches the earliest periods. Chapter 1 of Book VIII deals with, among other things, the comparison of animals and Man; in his childhood, Man is hardly more than an animal:

> If we have regard to the phenomena of childhood: in children may be observed the traces and seeds of what will one day be settled psychological habits, though psychologically a child hardly differs for the time being from an animal; so that one is quite justified in saying that, as regards man and animals, certain psychical qualities are identical. (Aristotle, *History of Animals*, Chapter 1, Book VIII, 343 BCE, D'Arcy Thompson translation, 1910a)

The potential form is specific to each species, the soul is the form put into action, or actualized in the substance, or in the physical world, its finality is the completion of the adult form. The *anima* confers to each animal its organization beginning from fecundation. The species are thus initially timeless forms and each species is quite distinct but the graduation is insensible:

> *Nature proceeds little by little from things lifeless to animal life in such a way that it is impossible to determine the exact line of demarcation*, nor on which side there of an intermediate form should lie. Thus, next after lifeless things in the upward scale comes the plant, and of plants one will differ from another as to its amount of apparent vitality; and, in a word, the whole genus of plants, whilst it is devoid of life as compared with an animal, is endowed with life as compared with other corporeal entities. (ibid.)

The formulation "Nature proceeds little by little from things lifeless to animal life" should read "the history of nature" is a system organized according to a graduation from mineral to animal. Some species are difficult to classify between plant and animal. The graduated scale is not a vertical succession of bars with *anthropos* at the top, but a ruler graduated horizontally with an observer. However, the graduation hides the boundary between degrees of organization.

1.2.3. *The generation of* anthropos*: the father as a model, the mother by default*

The question of the generation of the species is obviously "unthinkable" since the chromosomes that provide the indisputable argument to demonstrate a natural transmission are not observable. The reproduction of the "same" by the "same" raises the problem of memory, in some sense, of the animated construction of the organism after the male and female seeds have mixed. What is the memory of forms and the origin of movement? This memory is an idea, *eidos*, and movement is the activation of "a potential act", the *anima*. They are concepts, therefore pure abstractions. Like his master Plato, Aristotle conceived that his abstractions were inspired by a reality without materiality, without duration, a reality of ideas. This reality was first and thus the cause of the causes, it was metaphysical (the term being posthumous to his philosophical work) and it belonged to the celestial sphere which included the terrestrial sphere. What was the relation between the sperm, which is physical, and the celestial idea that corresponds to the species to be reproduced? The philosopher declined all possible hypotheses and concluded that the man's sperm must be in contact with the woman's menses in order for the seed to be formed. Only then were the form (*eidos* or idea) and the movement in power (*anima*, or soul) incarnated; the wall of the uterus was only the breeding ground that allowed the form to acquire its physical consistency with the movement put into action (kinetics).

Aristotle therefore distinguished the natural or physical plane, individually corruptible, and the timeless plane from the primary causes that order physics, which are a matter of metaphysics and not of direct observation of nature. They are the world of immutable causes, those that allow the second or physical causes. Philosophy is the knowledge of being, it is an ontology. The distinction between ontology and metaphysics did not exist in Aristotle's work.

Aristotle's concept of the soul was not the equivalent of the monotheistic religions that separated it from the body. With Aristotle, the body did not exist without the soul whose destiny was to be fulfilled in a physical reality whose limits it defined; it was considered potentially a space-time with limits and outlines that are internal as well as external, finite and, thus, predefined. The soul of the *pithekos* was different from that of Man. Movement, which is the most difficult observation to explain physically, was the *pneuma* of the soul. It is the breath within matter which animates it, so that the "shape" of the species unfolds, passing through a succession of changes from the embryo. In Aristotle's thought, it is primary causes that give each organized being its earthly purpose in an immutable cosmos; the philosopher did not conceive of generality between the different genera. The finality of a succession of animals from cold (invertebrates, reptiles) to warm (birds, mammals) with a gestation increasingly integrated into the body would have made no sense in a cosmos and on a planet without becoming. The central axis of the Earth is the axis of the world, not *anthropos*, even if its anatomical axis coincides with that of the Earth.

Aristotle was known for his patriarchal conception of nature:

> For by a male animal we mean that which generates in another, and by a female that which generates in itself; wherefore men apply these terms to the macrocosm also, naming Earth mother as being female, but addressing Heaven and the Sun and other like entities as fathers, as causing generation. Male and female differ in their essence by each having a separate ability or faculty, and anatomically by certain parts; essentially the male is that which is able to generate in another, as said above; the female is that which is able to generate in itself. (Aristotle, *On the Generation of Animals*, Book I, 343 BCE, Arthur Platt translation, 1910b)

The movement that animates matter was considered absent in the woman, but rather given to the sperm that transmits it to the menstrual period:

> Again, as the first efficient or moving cause, to which belong the definition and the form, is better and more divine in its nature than the material on which it works, it is better that the superior principle should be separated from the inferior. Therefore, wherever it is possible and so far as it is possible, the male is separated from the female. For the first principle of the movement, or efficient cause, whereby that which comes into being is male, is better and more divine than the material whereby it is female. The male, however, comes together and mingles with the female for the work of generation, because this is common to both (…). The animal is a body with Soul or life; the female always provides the material, the male that which fashions it. Thus while it is necessary for the female to provide a body and a material mass, it is not necessary for the male, because it is not within the work of art or the embryo that the tools or the maker must exist. While the body is from the female, it is the soul that is from the male, for the soul is the reality of a particular body. (ibid.)

The female was seen as a deformity that occurred during the development of what should have become a breeding male:

> For the female is, as it were, a mutilated male, and the catamenia[3] are semen, only not pure; for there is only one thing they have not in them, the principle of soul. For this reason, whenever a wind-egg is produced by any animal, the egg so forming has in it the parts of both sexes potentially, but has not the principle in question, so that it does not develop into a living creature, for this is introduced by the semen of the male. When such a principle has been imparted to the secretion of the female it becomes an embryo. (ibid.)

The female was therefore seen as an imperfect human being whose model was the man, and she was not pure in the sense that she lacked the essential: the gift of life, or the soul.

For example, she is less than the ideal form since she does not transmit the *eidos*; her finality (her place in nature or the meaning of her existence) is reduced to the nourishing receptacle of the small organism and to the transmission of the pure form, which would thus be that of the perfect man. In a way, the female is the placenta of the ideal or perfect man, while the male is his model which transmits the instructions to maintain himself in the physical world via generation.

In Aristotle's philosophy, since the human species was considered sexual, its purpose was ultimately the reproduction of the ideal model whose earthly image is the male, the female being only a derived state for the transmission of the species, having no soul in herself. For less elaborate organisms such as oysters or barnacles, which have no sex and are formed directly on rocks, Aristotle conceived spontaneous generation as dependent on the conditions of the environment. This is not uninteresting, since the natural environment conditions the formation of the animal, and if this environment disappears, the species disappears. It is not immortal:

> There is a considerable difficulty in understanding how the plant is formed out of the seed or any animal out of the semen (…). Either all the parts, as heart, lung, liver, eye, and all the rest, come into being together or in succession (…). That the former is not the fact is plain even to the senses, for some of the parts are clearly visible as already existing in the embryo while others are not; that it is not because of their being too small that they are not visible is clear, for the lung is of greater size than the heart, and yet appears later than the heart in the original development. (ibid.)

3 καταμηνια, *katamēnia*: menses.

Immediate generation by putrefaction concerns only relatively simple organisms such as oysters or certain insects:

> Some insects copulate and the offspring are produced from animals of the same name, just as with the sanguinea; such are the locusts, cicadae, spiders, wasps, and ants. Others unite indeed and generate; but the result is not a creature of the same kind, but only a scolex[4], and these insects do not come into being from animals but from putrefying matter, liquid or solid; such are fleas, flies, and cantharides. Others again are neither produced from animals nor unite with each other; such are gnats and many similar kinds. (ibid.)

The question of the beginning, of the first couple of each species, is considerable and tautological, with the circular problem of the chicken and the egg. That is why this aporia is overtaken by the idea of the animal-soul which finds its original soil from the wetland; the first movement, the first breath fertilizes the clay, from which emerges the male and female couple who each make their own seed, the archetype of Adam and Eve. Populations of the Near East developed an interpretation of nature and its history different from their Hellenic neighbors who occupied the western coast of Asia Minor. The cosmos had a beginning and, of all these populations, it is the Hebrew culture that has preserved it until the 21st century. The history of the cosmos is no longer natural, or physical; the cosmos does not last, it "does not become", it is not a temporality that unfolds by building itself. As with the Greeks, the history of the cosmos, of animals and Man on Earth, originates in being; the species remain separated with the same notion of a seed that bears their function or purpose. The difference is the being who becomes the creator in six stages:

> Let the land sprout with vegetation, every sort of seed-bearing plant, and trees that grow seed-bearing fruit. These seeds will then produce the kinds of plants and trees from which they came. Let the earth bring forth living creatures according to their kinds. (ibid.)

With Aristotle, the seed of Man thus had no other purpose than the terrestrial reproduction of his species' "eidos". Whereas, with Genesis 1, if all the animal finalities remain terrestrial and if each species is conceived for itself, it is different from the finality of the human species; it is not only terrestrial, but also of a cosmic essence. The first couple comes out of the breath of earthly dust, which is very close to Aristotelian philosophy. However, the human anatomy parallel to "the axis of the World" is replaced by the human anatomy based on a cosmic genesis finalized by the first creative cause of the cosmos. Man's finality is no longer simply the reproduction of his terrestrial species; it takes on a celestial dimension, inscribed in a

4 Larvae.

universal history inseparable from the first of the metaphysical causes. His destiny is not physical, or terrestrial, it becomes cosmic, and it therefore related to the metaphysics of first causes and the underlying ontology, the being-creator of the first causes. Man's destiny is to know himself "in the image, in the likeness" of his creator. This finality obviously does not refer to the verticality of the body which coincides with the axis of terrestrial gravity, nor to the wet brain and the two membranes which surround it. It refers to the necessity for human consciousness, at a given moment in one of the stories of humanity, to make sense in the animal and vegetable kingdom, and also on the terrestrial and cosmic scale. Populations are exposed to catastrophic earthquakes and floods, to epidemics. In certain cultures, humanity is in need of giving meaning to its finitude, corruption and death.

Wise men or astrological mages have sought to give meaning to the planet and its inhabitants, who share the solar system with other planets under the same celestial vault of a cosmos studded with stars. For pre-Socratic naturalist philosophers such as Alcmeon, or Socratic philosophers such as Aristotle, the question of the terrestrial finality of thought did not arise. The shift toward a teleological metaphysics, or finalized metaphysics, takes place smoothly; the finality of the human soul, which is reason or intellect, is therefore understood as the search for and understanding of first principles, the search for the essence of reason. The purpose of *anthropos* as the only conscious animal capable of reading the world with numbers is a cosmic ontology. It is nevertheless true that there can be no knowledge without the body, which is very earthly, with its sensory receptors, and a complex organ – the brain – to analyze the images or sensations that are captured and thus build its own image of a thinking or conscious body. This is the dilemma of the body–mind duality, of the transcendence or immanence of being in relation to the physical and psychic world.

Philo of Alexandria (20 BCE–45 CE) thus proceeded to syncretism with Jacob's ladder of souls and biblical finality. A thousand years would pass before Albert the Great and Thomas Aquinas (1224/5–1274) incorporated Christian doctrine, the cornerstone of which was belief in the resurrection of Yeshua (Jesus Christ), who died of the wounds resulting from the Roman torture of scourging and crucifixion. This school would impose itself in the rest of Europe from the 13th century by teaching the creation of the world in six periods and a scale of beings unrelated to the graduation of Aristotle's natural history of animals. Five centuries later, Georges Cuvier led a real struggle to free the natural history of animals from the literal reading of the Bible. However, while the idea of evolution gradually developed with the most daring minds, Cuvier did not abandon metaphysics and the separate creation of genera (which bring together several species).

In the 21st century, Aristotle's most obvious impact in primatology is the preservation of the name "anthropoid" to designate fossil species of monkeys that are dated to some 40 million years ago, as well as all their descendants, including humans, when they no longer exist on the planet. This inconsistency does not reflect a biblical finalism, it corresponds to the preservation of Aristotle's animal history. Never in the work of the Greek philosopher does nature read itself through a finality predefined by an archetype of which *anthropos* would be the fallen physical and therefore individually mortal version. *Anthropos* is taken as the origin of a comparison in the history of animals and therefore has no relation to Adam as the final cause of a damaged creation with mortal individuals.

Whether of Aristotelian or biblical origin, the philosophy of nature remains, however, devoid of the "immeasurable" that is duration. It remained so until the 20th century, when Henri Bergson (1859–1941) focused on the irreversible arrow of duration that unites two instants of a thought, or in the life of the same organism. Ilya Prigogine (1917–2003) gave it a scientific formalism with the dissipative structures rewarded by the Nobel Prize in Chemistry in 1978.

2

From Aristotle to the 16th Century: The Eclipse of Science

2.1. Introduction

Georges Cuvier's epistemological research is particularly enlightening in evaluating the sum of scientific knowledge and concepts acquired at the height of the French Revolution of 1789, and thus on the eve of the first theoretical formulation of the natural kinship relationships between extinct and present-day species. The great naturalist concluded that, between Aristotle and the 17th century, natural history shows "almost nothing" new.

We will therefore follow the major stages in the way in which knowledge of physics, chemistry and biology was acquired up to the generation of Georges Cuvier, pushing metaphysics ever further into the confines of the human psyche. Then, we will follow knowledge of the comparative anatomy of primates and Man (i.e. the Homo genus) in particular, ending with the major conceptual stages of the 19th century. The major references are Georges Cuvier, Henri Daudin (1926–1927), Robert Baccou (1951), Emile Guyenot (1957) and Bernard Balan (1979), supplemented by more recent work.

2.2. Comparative anatomy of apes and humans from Aristotle to Galen

2.2.1. *The Museum of Alexandria*

2.2.1.1. *Followers of Aristotle*

Knowledge of human anatomy and physiology varied according to the various Greek schools and the quality of their transmission (Annoni & Barras 1993). Alcmaeon of Crotone (ca. 490–430 BCE) is said to have incorporated the use of dissection and perhaps vivisection, according to Calcidius (4th century CE), giving an elaborate description and interpretation of the organ of sight and the function of vision:

> It is therefore necessary to demonstrate the nature of the eye: there are two narrow ducts which, starting from the brain, where the master and directing part of the soul is located, connect it to the eye sockets: it is through these that the vital breath circulates. These ducts, which, starting from a single point of departure, merge over a certain distance in the deepest part of the forehead, separate into two branches to join the eye sockets, following an oblique line which is that of the eyebrows. And, after detours, they culminate in globes which contain the vital humor within tunics and which are protected by the eyelids, hence their name of orbs. What allows us to know that the origin of these channels that transmit the light is unique, is of course, the section (…). (Crotone, in Annoni & Barras, "La découpe du corps humain et ses justifications dans l'antiquité", 1993, author's translation)

In 290 BCE, Ptolemy I Soter, king of Egypt and follower of Aristotle, had a Mouseion built in his imperial capital Alexandria in order to gather all knowledge, provide teaching, develop research and preserve the most complete library. Two Greek physicians founded the School of Medicine and Anatomy of Alexandria there, Herophilos (330–250 BCE) from Chalcedon (between the Marmara and Black Seas) and Erasistratus (310–250 BCE) from Ceos (an island in the Aegean Sea). Herophilos appears to be one of the first physicians concerned with understanding the normal physiology of man. He resorted to animal and human dissection. His work was recorded in nine treatises on anatomy that were destroyed in the library fire in 47 BCE. He described the brain surrounded by meninges, the venous sinuses at the base of the skull and the cranial nerves.

> He analyzed the nature of the pulse and elucidated the structure of many other organs. (Cuvier, *Histoire des sciences naturelles*, 1841–1845, author's translation)

His colleague Erasistratus of Ceos is said to have performed nearly 600 dissections. He described the organization of the brain with the cerebellum and the two hemispheres of the brain, showing how the nerves converge toward the central nervous system, and thus managed to distinguish between motor and sensory nerves. He understood the role of the cerebellum in the motor coordination of movements and assumed a correlation between the number of cerebral convolutions and intelligence, like Alcmaeon, who located its center in the brain. He also described the valves of the heart, distinguishing between arteries that have a pulse and veins that are inert. The circulation of blood in the vascular network was obviously one of the major pre-occupations, though physiological relations with the brain still challenged these physiologists, those between the vascularized meninges, the venous sinuses located at the base of the skull, the pulse and the cardiac rhythm. The war wounded, who required surgery, must have shown blood pressure in the vessels; the heartbeat was most likely observed *in vivo*.

There was no shortage of monkeys in Egypt to complete the anatomical comparisons in the Alexandria Museum, especially since they were eaten or mummified. Perhaps comparisons between their brains and those of humans allowed Erasistratus to suppose a relationship between the number of convolutions and intelligence, even though monkeys also had hands and used them skillfully.

No written record of an anatomical comparison between the bony envelope of the brain and spinal cord of apes and humans is known, perhaps due to the destruction of anatomical treatises, whereas in humans it shows the verticality of the entire cerebro-spinal nervous system from the center of the base to the last vertebra. Since blood passes through the heart, the understanding of the blood circulation of the human compared to that of the ape, or another quadruped, is understood by its position relative to the brain, therefore relative to verticality. The amount of movement of the blood fluid needs more energy to reach the sense organs of the head and the encephalon, fighting against the pull of gravity. It would be surprising if the two doctors in Alexandria did not consider this difference, aware of taking the ape as a model of human physiology and blood circulation.

Knowledge of physiology was also gained through vivisection of those condemned to death (Byl 1997). The torture was flogging in public, to make an impression on citizens, and it differed between Alexandrian and Egyptian prisoners (Taubenschlag 1948). The law clearly allowed for medical care of those tortured by such doctors before the end of life. The Egyptians treated wounds with aloe oil, a method which allows the blood to clot and heal scrapes and open wounds. It was in the Pharaoh's interest that his doctors find the best treatment for those greatly wounded in war. It seems legitimate to wonder whether doctors such as Herophilos and Erasistratus were not practicing experimental treatments on tortured people in accordance with Hippocrates' deontology, rather than lending themselves to experiments described as cruel by a 4th century Christian with no evidence to support it.

2.2.1.2. *Galen, or Galenus: extrapolating the human being from the anatomy of the ape*

Galen, born in 129 CE in Pergamon (Asia Minor or western Turkey), was the last of the great Greek physicians of Antiquity along with Hippocrates. He lived during the apogee of the Roman Empire and was the physician of his ruler, Marcus Aurelius.

> The Greeks were so ignorant of anatomical science that they did not even know the osteology of man, for Galen himself made the trip to Egypt. (Cuvier, *Histoire des sciences naturelles*, 1841–1845, author's translation)

In the meantime, Roman law had forbidden the study of the human corpse. Galen therefore learned dissections, experiments and surgical procedures on animals and apes in particular. Back in his homeland, he was appointed "doctor of gladiators" by the emperor Marcus Aurelius, with the instruction to save as many of them as possible. He cured them by applying the treatments taught by the Alexandria School of Medicine with oils, and they were a success compared to his predecessors, who poured hot water on their wounds. He took advantage of these operations to learn physiology and advised dissecting apes to become familiar with surgical gestures.

Galen is said to have seen only two human skeletons but Cuvier wondered to what extent he would not have had skeletons of apes and humans at his disposal, given the very precise description of the specific characteristics that distinguish them. In fact, he noted an excrescence on the human scapula at the level of its articulation with the clavicle (the acromion), absent in the ape, and he interpreted it as a necessary reinforcement for the consolidation of the skeletal framework. It is likely that he related this difference to human verticality as opposed to the quadrupedal carriage of the ape, with its lateral rather than dorsal shoulder blades. When Galen reached Rome, where his notoriety was finally established thanks to his mastery of *in vivo* experiments, he demonstrated to the medical profession that it was indeed physiology that allows the functioning of the senses as well as articulated language. For this demonstration, he used goats and cut their nerves, causing paralysis of the phonation muscles. He demonstrated that breath comes from the exhalation of air through the lungs when the thoracic cover retracts and showed how sound is caused by its passage through the vocal cords.

Physiology thus showed the unity between organs and their functions, as well as between organs and those senses shared with animals and those specific to humans, such as articulated language. Galen's lectures noted by stenographers have not been preserved; we do not know if he had the idea of correlating the verticality of the larynx, its musculature and innervation, with the articulated language absent from quadrupeds, or if he had noticed that speech is transmitted by the exhaled breath, while the animal cry occurs inversely during inspiration. It is almost impossible for human physiology to express speech while breathing in, which seems to be correlated with the verticality of the airways. The relationship between air and blood was confused, as the physical and physiological relationships between the heart, lungs, upper airways and vascular network were still not elucidated. By analogy with breathing, Galen reasoned in terms of blood flow and reflux, distinguishing veins and arteries that did not contain the same quality of blood. However, his model was particularly complicated, with the liver as the organ that makes blood and a crossing of the blood through the wall of the heart separating the two ventricles. The place and date of his death are approximate; he is estimated to have died at the age of 89 years at the beginning of the 3rd century in 216, either in Rome or after returning to his native land in Pergamon.

2.3. Decadence and rebirth of natural philosophy and human anatomy

Aristotle's geocentrism was developed by the last of the Greek philosophers of Alexandria, Claudius Ptolemy (100–168 CE), followed by an eclipse of science that lasted more than 1,000 years following the decline of the Roman Empire. Rome was no longer an imperial city; Byzantium became the epicenter of a Christianization of Europe since the emperor Constantine converted to the new religion, who chose in 324 CE to make it the eponymous imperial city, Constantinople. Galen's medicine and anatomy disappeared from Western Europe. His body of knowledge continued to be transmitted in Constantinople and Alexandria and was read by the Persian physicians al-Razi (865–925) and Ibn Sina (Avicenna, 980–1037), who translated it into Arabic. His medical corpus returned to Western Europe in the form of Greek, Byzantine and Arabic texts.

2.3.1. *Albertus Magnus, the Aristotle of a reborn Europe*

Albertus Magnus (Albert the Great) was born in Swabia (Bavaria) between 1193 and 1206 and died in Cologne (Rhineland) in 1280. After studying literature (Latin, Greek, Hebrew and Arabic) and medicine at the Universities of Venice and Padua, he entered the religious order of the Dominicans and devoted himself to theology. From 1241, he taught at the University of Paris and discovered the Latin version of Aristotle's work, translated by Michael Scot (1175, died around 1232), as well as Ibn Sina's treatises on medicine. He embarked on a very ambitious project aimed at the rediscovery of the philosophy of natural sciences and its teaching in the universities of a Christianized Europe.

Albertus set upon the realization of a work that he wanted to be comparable to that of Aristotle by complementing it with the fauna, flora and minerals of a Europe unknown to the Greek philosopher. He thus wrote *De animalibus* in 1258, an encyclopedia which includes 19 books taking up Aristotle's classification in the form of a commentary, rectifying it when necessary, to which were added seven books on his own research. He also carried out investigations with those very knowledgeable in nature such as hunters, whalers and falconers. The 26 books were copied and illustrated in the 14th century:

> While the manuscript thus takes the appearance of a charming bestiary, the illustrations, following the text of Aristotle, do not fail to also approach the study of the human species: the human body, its organs, generation and childbirth are illustrated with an astonishing medical precision. Finally, the artist, with a concern for scientific realism, has sought to represent the academic world on various occasions, mixing scenes of active observation with scenes of lectures during which the master transmits his knowledge to an audience of attentive students. (Postec, "Un exemplaire singulier du De animalibus d'Albert le Grand et son illustration", 2014, author's translation)

However, he had to find a compatibility between Christian doctrine, or scholasticism, and Greek philosophers. Comparing Aristotle and Plato, he opted for the former. Christian Aristotelianism thus generated a Christian anthropogenesis, distinguishing, on the one hand, Judeo-Christian theology (the Torah and the Gospels), and, on the other hand, natural philosophy with metaphysics and sciences based on observation as well as on experimentation, which seemed indispensable to him. This concordism claims a universal intelligibility of the origins of Man and animals. In doing so, Albertus Magnus added to the natural history of animals, including Man, an event considered supernatural and the most significant in the genesis of the sublunar world: Yeshua (Hebrew name, Iesos in Greek and Iesus in Latin) had died twice, first as a result of his injuries and again after a few weeks.

Scholasticism was a religious doctrine built on dogma and not on reason. The independence of the metaphysicists and physicists of Antiquity was thus muzzled by the unassailable and incontestable dogma of the creation ending with the resurrection of Yeshua. Understanding the place of plants, animals and humans and their pathology in nature therefore resulted in more than six centuries of struggle against a more or less rigid dogmatism, depending on the political and ideological interests of the rulers and their relationship with the Papacy.

Treatises on medicine and surgery were translated into Latin in the 11th and 12th centuries by scholars who made them available in convents. It was the religious and clerics who preserved the texts and copied them without any research to verify them. However, in 1163 the Catholic Council of Tours forbade surgical acts, considering that blood was to be abhorred because of Yeshua's torture. In 1215, the Fourth Lateran Council forbade priests to perform surgery. The relationship between the sick body and blood was censored. The church soon lost control of the knowledge of human anatomy and physiology and it was lay initiatives that took up the challenge.

2.3.2. *The first lay schools of medicine in Europe in the 11th and 12th centuries*

During the 9th century, lay doctors from the city of Salerno in southern Italy were allowed to train students. This was the first lay school of medicine in Europe. In France, the first school was founded at the end of the 12th century in Montpellier in 1180. Teaching was reviving the school of Alexandria, thanks to Greek, Byzantine and Arabic texts. The teachers were content to read the texts of Galen, while popular knowledge transmitted the therapeutic gesture and the preparation of medications by plants, orally and by initiation. This distinction between the doctor who professed but did not treat, the barber-surgeon who operated and the pharmacist who put his hand to the work was legislated in the 13th century. Frederick II, at the head of the Holy Roman Empire, intended to reform the training of doctors and supported the arts and sciences by welcoming many scholars. As early as 1231, he granted specific jurisdiction to the medical school of Salerno, which could incorporate itself as a corporation of masters and students and obtain the status of university, with the right to train surgeons and teach dissection. The Pope excommunicated him twice. Never mind: the University of Salerno trained doctors and surgeons who scattered throughout Italy and in turn founded guilds that allowed them to create their own university, especially in Bologna, south of the Po plain, where the first human dissections could begin.

2.3.2.1. *From the barber-surgeon to the physician-anatomist*

Thaddeus of Florence (circa 1215–1295)

Thaddeus established a doctorate in medicine between 1260 and 1270 in Bologna and thus founded the new University of Medicine. Observation, as well as collection of materials for lectures, was a menial and messy task. The doctor was a literate (he learned Latin, Greek, Hebrew and Arabic), he was a "reader" from the pulpit who read the works of the masters and remained distant from his subject. He was not yet a "savant" nor a physician, literally speaking, as he did not try to understand and observe by himself. Courses on dissection were entrusted to the barber-surgeon who practiced the method on dogs, or even on the corpses of hanged men recovered from the gallows when circumstances allowed it.

Mondino de' Liuzzi (Mundinus) (1270–1326)

Also a Florentine, Mondino succeeded Thaddeus in 1321 after writing a textbook, *Anathomia* or *De anatome*, for his students. In it, he describes for the first time, but without illustration, the dissection of two corpses of women. The work is an affirmation of the independence of medical universities from scholasticism. The textbook was distributed and copied in 1315 and was published posthumously in several editions. The first one appeared in 1478 in Padua and then in 1493 in Leipzig with an illustration showing the professor reading from the pulpit and the barber-surgeon dissecting (Mondino 1551).

More than a century of teaching had gone by without any perceptible change. Then the end of the 15th century and the beginning of the 16th century heralded a renewal of interest in the human sciences in the universities, again motivated by the need to free themselves from scholasticism. It was the end of the Middle Ages with the first tremors of the Renaissance.

Allesandro Achillini (Magnus Achillinus) (1463–1512)

Born in Bologna and dying in the same city, an independent spirit and curious about everything, Achillini continued his studies at the University of Bologna and the University of Padua, where he taught anatomy and philosophy. As was customary, he published commentaries on the texts of the Ancients, the soul according to Ibn Rushd (Averroes), the "wonders of India" described by Alexander and on Aristotle, which earned him the reputation of being the "second Aristotle". He was one of the first academics to dissect human bodies and to note the errors of Galen. He published his own observations in medicine and human anatomy, as in *Anatomicæ annotationes* (1520); but, without illustrations, his observations were of very limited use due to the lack of visual reference for the surgeon.

2.3.2.2. *A turning point toward modernity: Jacopo Berengario, feminist and libertine physician*

Jacopo Berengario, born in 1457 in Carpi – a town near Modena in the Po plain – marked the turning point of modernity; he was a great pedagog and an avant garde humanist. He began by following the education of his father, Faustini de Barigatiis, a renowned barber-surgeon. He mastered hysterectomies, consisting of removing the uterine body, the cervix, the tubes and the ovaries. Jacopo Berengario was therefore familiar with Mondinus's book for students and was fully aware of doctors' visual ignorance of the female body and reproductive organs. He decided to become a physician and followed Achillini's dissection courses at the University of Bologna, where he obtained his doctorate in 1489, at the age of 32 years. After a few travels between Carpi and Rome, where Pope Clement VII (Julius de Medici, 1478–1534) offered him the opportunity to become his doctor, he returned to Bologna. He was appointed surgeon and lecturer (professor) at the University of Medicine, where he taught from 1502 until 1527, when he was forced to give up his professorship for a reason that remained obscure and still arouses curiosity.

Like his predecessor, he lent himself to the practice of commentary. It was not yet a question of describing dissection for its own sake, but of arguing a text by reinforcing it with dissections, in this case, the text by Mundinus (French 1985). Jacopo dissected more than a 100 corpses, including women, and printed his first

"Commentary on Mondino" in 1514. He realized the abstruse nature of this method for novices, it being purely literary. The professor's reading remained incomprehensible to the demonstrator, as well as to the student, who could not properly imagine what they had never seen.

Jacopo Berengario then developed a method of teaching his own published and properly illustrated observations.

Human dissection remained a practice subject to controversy, situated between the interest in observing an inert body that was no longer related to physiology and the surgical act that doctors were reluctant to perform because it was similar to vivisection. Jacopo therefore intended to publish a commentary on this subject, all the more so as he planned to illustrate his dissections in a very original way, aware of the paradox: *since the aim was to treat a pathology, teaching from the dead in order to understand their physiology, which was absent, required reconstructing the living appearance based upon the surgical acts performed on the patient.*

It was enough for him to make hypotheses and test them, as long as the experiment did not undermine his integrity and dignity.

In 1521, Berengario put into print a second commentary of 1,000 pages with descriptions drawn by his own hand. It was the living body that he restored, for the first time in the entire history of medicine. The dilemma between the anatomist who dissected and the physician-surgeon who treated was overcome by the knowledge of anatomy and organs, which was essential for operating without risk.

The Commentaria cum amplissimus additionibus super anatomiam Mundini *("Commentary with More Documents on the Anatomy of Mundinus") (Berengario 1521) became the very first treatise on human anatomy with reproduced illustrations of the observation of the skeleton, superficial muscles and the reproductive system of women.*

These are 19 particularly avant garde plates of unprecedented free thought. Engraved on wood, his illustrations are simple but precise. The sensation of volume of the body is rendered by pencil strokes that orient the shadows.

The woman is depicted standing, naked and dissected, but in a living posture, not enclosed within domestic walls (Plate CCXXVII in Berengario 1521). She is looking at the reader, she is the teacher lifting the veil with one hand, pointing to the genital apparatus, and with the other hand she is pointing to the uterus that has come out of her abdomen, her foot resting on the books. Another plate shows her as a work of art, a living sculpture of nature on its pedestal that takes the liberty of showing its organic interiority. Jacopo Berengario was the spokesperson for the woman who taught the student about the anatomy of her reproductive organs, telling him that knowledge is in her body and not in books. The woman's body returns to her, it is no longer the regard of Aristotle that makes her a soulless degenerate male or the Eve of the scholastics that caused the fall of Adam.

As for the male body, he represented it naked with muscles and then, in the same environment, he drew the skeleton out of the tomb in a living posture, without anything indicating sex. The two feet raised on the tombstone, the legs are in a position that indicates movement, while the face turned toward the distant city gives perspective for the very first time in an illustration. Each hand is busy showing the hidden sides of the skull, the left holds it like a world map in cranial view and the right in a slightly three-quarter right lateral view. The scene shows, as it were, the living body in transparency stripped of its flesh (Berengario 1521, plates CCCCCXIX to CCCCCXXII).

Five plates depict a standing naked man lifting his skin to show his stomach muscles. Most unexpected was his choice to place the naked man's body back on a cross with his muscles nailed to his hands and feet, but he is not bleeding, since he is dead. The plates are visible online (Berengario 1521, plates CCXXVII and CCCCCXX).

Basically, it was the most explicit stand against the prohibitions of the Councils of Tours and Lateran on dissecting corpses. Blood was not drawn, it does not flow. The man does not look at the reader as opposed to the woman, so the body is dead. This is the only corpse represented. It does not refer to Yeshua, but to the ancient masters, such as Galen, Herophilos and Erasistratus, strongly disputed by Celsus (37–29 BCE) and John of Alexandria (bishop of Alexandria in 487 BCE), who discussed the advantages and disadvantages of human vivisection, practiced on those condemned to flogging or crucifixion, which were reserved for men.

Berengario necessarily commented on this at the beginning of his treatise, explaining that the "subject of this book" is "the human body, living or dead".

He insisted on the distinction between the surgical act that cures and vivisection that kills, recognizing that the surgeon can make mistakes, but insisting that under no circumstances should he operate on the vital parts of a human being such as the heart, brain, kidneys and reproductive organs. The crucified man drawn is perforated only at both wrists and both feet, his vital organs are not affected. The language thus communicates on several levels.

These drawings have the merit of showing the complexity of the organs necessary for the reproduction and intrauterine formation of the human body. They show great daring in imposing a naturalist and intelligent view of man and woman, free from Aristotelian patriarchy and the guilt of scholasticism. These plates have the immense advantage of literally "throwing light" on the obscure and hidden areas of the body, helping a speculative medicine emerge from an obscurantism more than 1,000 years old (Cazes 2019) by breaking the prejudices conveyed by many doctors, such as Gerolamo Cardano, remaining in ignorance of female physiology and peddling sexist prejudices about the fault of mothers for fetal malformations.

Jacopo Berengario took great liberties, knowing that the University and its teaching were not accountable to the Catholic Church. Furthermore, no drawing violates the dignity of a woman and a crucified man; the book is pedagogical, and nudity is human nature.

In 1522, he had a new treatise published with wood plates cut in Bologna by the workshop of Hugo da Carpi (1455–1523); they are of much better quality, with embellishments around the figures printed in the *Anatomia Carpi: Igagoges breves* ("A Short Introduction to Anatomy") (Berengario 1522). The dissections criticize Galen's Treatise on Anatomy; the anatomical plates show in particular the tightness of the interventricular septum of the heart. They thus allowed the other professors to verify for themselves the relevance of his observations. New plates were added with the very first impression of an anatomical drawing of the human spine seen from the side and back (plates in Berengario). The vertebrae are numbered with the relief effect, as well as their two ends with the atlas and the pelvis composed of the sacrum, the caudal bone (the coccyx) and the right coxal bone. The brain is drawn from above before and after its dissection, showing the ventricles; three plates show the dissected heart with the inter-ventricular septum and the reproductive system of the woman is drawn.

The head is better analyzed than in Mondinus's work, both osteologically and organically, with a description of the different parts of the brain and skull. Antoine Portal gave a good recension (Portal 1770).

Berengario described for the first time the basilar part of the occipital bone, which corresponds to the verticalized part of the base of the human skull, and which is extended by the spine. He observed the larynx and described the five cartilages; he noted the presence of sinuses (air-filled cavities) in the frontal squama above the orbits and in the bones of the base of the skull, the ethmoid (behind the nose bones) and the body of the sphenoid that occupies the center of the base, noting that these cavities can communicate with each other. In the petrous pyramids, Berengario describes the middle and inner ear, with the eardrum separating them, as well as the two small "ossicles" of the ear, already described by his professor Achillini, but not named. He described them as "hammer" and "anvil" (the stirrup is missing) and also noted the presence of the "snail" (cochlea).

His descriptions are not exhaustive since the three semicircular channels – which give the three dimensions of Euclidean space – are absent.

Four years after the publication of the second work, in 1526, the Court of Inquisition sued Berengario, who was forced to surrender his professorship, having just enough time to take refuge with Duke Alfonso I d'Este in Ferrara, near Bologna. The latter assured him his protection by welcoming him as a doctor at his court. Jacopo Berengario died 3 years later, in 1530, at the age of 73 years. However, no sentence was pronounced, which fueled the worst of the speculations. The rumor spread by the famous doctor Fallope, his competitor at the University of Padua, claimed that the professor from Bologna had performed vivisection on two syphilitic Spanish twins.

Two hundred and fifty years later, Daniel de La Roche and Philippe Petit-Radel doubted the veracity of the "vulgar rumor", as they put it, because if it had been a question of manslaughter, the Court of Inquisition would probably not have let him leave Bologna. The circumstances that prevailed at the end of his career still remain an enigma for contemporary historians. "On his return to Bologna, Berengario lost his position as Lecturer in surgery and, for a still unknown reason, left the city in haste" (Parent 2019).

The reason is not so unknown. Antoine Portal (1770) evoked this drama by referring to an article "on generation", in other words, on the reproductive organs of women. As Jacopo Berengario's text cannot be incriminated, it is toward iconography that the motive for the accusation should be sought. Indeed, the comparison of the illustrations between the two editions of 1521 and 1522 is revealing. All the 1521 plates are preserved in the 1522 version except for the three living women. They were replaced by a single illustration that literally erased the "feminism" of the physician. Only one plate remains. The naked body drawn by Hugo de Carpi reflects the ambiguity of the body of the crucified man of 1521. This disappeared, replaced by the woman with the absent gaze, who no longer provides her teaching. The sheet of the dissection table is thrown on a chair on which she is slumped, eyes closed, arms suspended under the armpits by the back of the chair, hands inert, feet hanging in the void, the books have disappeared, the legs are spread apart. The woman is no longer the mistress of her body and of the unveiling, it is the male artist who stages his model in the walls of the house and who has placed the sheet on the chair; she is returned to her domestic function. The only person authorized to unveil is the engraver, more voyeur than observer. Is she dead?

By comparing the drawings, a posture emerges of the body of a woman, whether dead or alive, shameless, in a lascivious and passive position. Commentators have not failed to note the intrusion of an erotic representation (Cazes 2019) that breaks with the spirit of the physician who refused to reduce the woman to a sexual object.

The disappearance of the image of the living woman, who looks at the reader and gives him his lesson, is all the more challenging since the posthumous edition of Mondinus (ibid.), later than that of Berengario, takes up the theme of the woman who unveils her reproductive organs, as in Charles Estienne, *La Dissection des parties du corps humain* (1546), where the eyes are closed but the dynamic posture means that she is very much alive.

Jacopo Berengario revolutionized medicine by inventing biological anthropology, which became a natural science thanks to the illustrations of the observations and comments of the researcher-teacher. With drawing, measurements became possible. Berengario appeared not only as a humanist but also as a feminist *avant la lettre*. As such, he was one of the most remarkable anatomists of the first generation of "moderns", or "humanists", opposed to the defenders of a medieval culture.

2.3.3. *Instant of grace: Leonardo da Vinci, from the elusive movement to the restitution of the soul*

Leonardo di Ser Piero da Vinci (Leonardo the son of Master Piero da Vinci) was born in 1452 in Anchiano, a hamlet of Vinci in Tuscany, not far from Florence. He died in 1519 in Amboise in *la douce France* on the banks of the Loire River. His works herald the spirit of the Renaissance. His drawings and sketches are preserved in several libraries and include:

– The Codex Arundel of the British Museum (1480–1518);

– The Codex Windsor at Royal Windsor Castle (1485–1490);

– The Codex Atlanticus of the Ambrosian Library in Milan (1485);

– The Codex Forster of the Victoria and Albert Museum (1487–1505);

– The fourteen manuscripts of the *Institut de France* (1487–1508) ;

– The Codex on the Flight of the Birds of the Royal Library of Turin (1505);

– The two Madrid Codices in the National Library (1490 and 1499 and 1503 and 1505);

– The Codex Leicester from the Bill Gates Collection in Seattle (1508–1510);

The codices have been digitized and are available online.

The whole of his work surpasses his contemporaries by the extent of the field of his exploration and differs radically from them, both in the method and in the quality and quantity of the transcription and visual transmission of his observations (Arasse 1997). Leonardo da Vinci was a humanist, his gaze turned to nature to decipher it, and it is the study of architecture, anatomy, neurology and human physiology that allows us to approach this capacity of the human to know how to interpret the world, to understand natural history and great principles. The breadth and multidisciplinarity of his work remain unequalled by the richness of his imagination as an engineer, geometrician and mechanic, by his anatomical precision and physiological graphic essays. He remains to this day a thinker that instructs us on the meaning of research and the art of its transmission without giving anything away to emotion (Vinci 2019).

2.3.3.1. *The learning years: intuitive exploration of movement*

Leonardo was not literate. Born out of wedlock, higher education was closed to him. He was modestly raised in the country during his first five years by his maternal family. His father was a notary from a wealthy and reputable family. When Leonardo was six, he took him in and provided for a short course of schooling to teach him to write and read in a Tuscan dialect. He would have acquired the basics of algebra to calculate with an abacus and some basics in the philosophy of nature. Above, however, all he possessed a remarkable gift that went hand in hand with an extraordinary thirst for knowledge and understanding. This gift was that of observation and drawing. He read Pliny the Elder's "Natural History" in Italian, a reading that was decisive. Not knowing how to translate Latin, Greek, Hebrew or Arabic, he did not have the rich vocabulary of a poet, engineer or doctor, which delayed his access to knowledge through texts.

Gifted with a sense of analysis, reflection with meditation and transcription through manual drawing, his work became that of such a thinker as humanity has rarely known. Rarely has such a keen observer of nature come so close to the link that unites the singular psyche of the human to the laws of physics and to a physiology verticalized from the center of the base of the skull; he seemed to be the only one for 500 years. He could not help but understand, observe and geometrize the dynamics of becoming upright, all the more comforted in his intimate conviction that in the human being is woven a universal history shared with all living beings down to the smallest elementary particle of the organism.

Leonardo's "secondary school" was fruitful precisely because of the freedom he had to undertake, innovate and explore in the artist's studio of Andrea del Verrocchio (1435–1488), one of the most prominent in Florence. This studio was a true multidisciplinary laboratory. It broke with the codes of Byzantine art: sculptures, copper engravings, paintings on wood or canvas and goldsmith's work left the studio according to criteria that aimed at a return to reality and natural figuration, be it the characters of the Bible or the venerable personages in their ducal palaces. The eye and the gesture of the craftsman reclaimed their creative and inventive rights from the word and writing. To the drowsiness of the spirit of the university responded the great breathing of the people who grew up in the country hamlets, with the miller and the moving gears of their watermill or windmill that pressed the oil and ground the grain, the blacksmith and their incandescent forge that melted the iron, who twisted it and beat it with the right gestures at the right temperature, the carpenter with their compasses and plumb line for the right measure, the right angle and the right load, the barber-surgeon with their scalpels, the midwife delivering the baby, retrieving the placenta, cutting the umbilical cord and tying it with her nimble hands forming the navel.

The internal organization of the human body was not well known; diseases were cured by plants or concoctions of mineral powder and venoms of various animals. The organization of the animal was of interest only for profit or for the food trades, and not out of curiosity. Its study concerned the horse, a draught animal in the countryside, a prestigious animal in the conquests and the wars which were worth so much nobility to their victors.

In Verrocchio's workshop, Leonardo da Vinci discovered the art of drawing and painting. He, who sought to be as close as possible to life, discovered the means of his expression: movement, the sound of the string or wind instrument, the changing tones of colors with the light that followed the movement of the Sun and the Moon, and the reflection of shapes, movements and colors on the surface of puddles, ponds and lakes. He was intuitive and not a mathematician, unlike his young contemporary Nicolas Copernicus with whom he had a cryptic correspondence. The young astronomer scrutinized with the naked eye the sky and the movement of the Moon and the stars with a measuring instrument, but the method of calculation was complex and da Vinci did not master it. He knew algebra (counting and enumerating) but not arithmetic (calculating equivalences between distances and movements) and even less trigonometry (ratios between angles and lengths), but it did not matter. He had its own method. At the age of 26 years, his talent was recognized, he received orders and opened his own workshop in Florence.

It would be inappropriate to consider observers of nature as physicists, as physics did not yet exist. Knowledge was based on generations of experience and observation, through repetition, efficiency and reasoning. Gravity was an observation, it was an *impetus*, i.e. a causal agent that gave a quantity of movement (force) to a solid, liquid or even gaseous material, but it was not measured and nobody knew where and when this quantity of movement, which never stopped, began in time and space (this was then the domain of metaphysicists). For Leonardo da Vinci, this *impetus* was a permanent flow, these were continuous vibrations that propagated, which were reflected from the sunlight reciprocally from the Moon to the Earth and between the other planets.

However, mathematicians did not yet know how to quantify the movement of a wheel on its axis through mathematics – this science was kinematics – nor could they quantify the speed and acceleration of a waterfall – this science was kinetics – nor could they quantify the energy necessary to accelerate or slow down the mill wheel – this science was dynamics, or the sciences of force.

This is what Leonardo was looking for, so he invented a geometric mathematics that did not measure; it correlated ratios or proportions, but without putting them in the form of an equation. These proportions and ratios made it possible to geometrize perspective but not to explain, in particular, the physical process of any movement. His vocabulary was familiar with terms such as energy, motion, weight and force. However, of all the "sources" that animate reality and connect humanity to nature at all levels of organization, the one that most fulfilled his need for knowledge and coherence was obviously the nature of light.

Without light, there can be no observer and interpreter of nature. What was the light that transmits to the back of the eye, the color and size of objects that vary according to lighting and distance? What happens to it from the back of the eye to the brain? These phenomena in animals and humans were admirable. Thus, his studies sought to describe the functioning of vision and his talent as a painter allowed him, through his mastery of colors, to experience the indefinite passage from light to shadow (*sfumato*).

Leonardo da Vinci gradually tried to understand how the organism "worked" but without ever being able to deduce an empirical mathematical law. As such, we would not say in the 21st century that he was a scientist, since his work was not a treatise with calculations, statistics and equations. It was neither a scientific theory nor a system of thought that described physical properties by mathematical rules, those that would allow us to build a bridge based on calculations. Bridges were built by relying on proportions and empirically, through the experience of masons, carpenters who observed the behavior of materials under stress without needing to know their physical properties.

The artist Leonardo was an explorer, an observer and a thinker who allows us to see, to discover and to reflect with him on what his glance and his reflection reveal to us about the invisible side of life, of movement, of the animal and of what is seen in the human that is not seen in other animals. The innumerable washes or pen drawings are the studies of a genius researcher gifted at restoring perspective on a flat surface and through sensory experimentation. He absorbed reality through his senses and restored it through drawing, while meditating on the principles that govern its balance and functioning as the line turns this way and that. He sought to visualize all kinds of processes, with the need to distinguish movement, energy, force and mass to give them a physical appearance, visible through drawing. In 1485, he planned the publication of a treatise as he wrote on a sheet of the Codex Atlanticus *un mio trattato di moto locale: forza e peso*[1] (Vinci, *Codex Atlanticus*, 1485, see online, p. 978).

He wanted to make visible the natural forces that underlie everything, the turbulence of the river waters, the circulation of blood in the two ventricles of the heart and generation by the functioning of the sexual organs seen in cross-section. His intuitive understanding of natural movements such as that of a waterfall, or the flight of a bird, which he geometrized in his own way by means of spirals, allowed him to move from drawing to three-dimensional models, articulated and moved by moving elements such as water, air currents or animal and human force. Some of them were doomed to failure for lack of a possible mathematical formalization between mass, terrestrial attraction and aerodynamics (the "flying machines").

He used letters as markers on a graph and numbers to count segments. It was a mode of expression by linear quantities that he measured when necessary. He compared the A–B segment to the C–D segment, it was the ratio that mattered. He found a "growth law" for trees, namely, a constant between the diameter of a trunk and that of the branches (Vinci, *Manuscrit M de la Bibliothèque de l'Institut de France*, folio 78 v, 1487–1508). The fall of a body and its acceleration, which imbues it with power – visible by the impact on the ground – were geometrized in the form of a narrow pyramid with segments whose width widens and height decreases as the object approaches the ground; but he did not infer an empirical law from this.

He was unable to calculate (at the age of 50 years, he could not apply multiplication), but the method of segments allowed him to reason. No matter how far apart the two segments were from each other, their ratio must not change. Of course, as far as possible on the horizon, the segments merge, and beyond, the whole forms a point and then disappears. The capacity of the eye to distinguish at great distances is limited. To see toward the infinitely far (the stars) and toward the infinitely small would require magnifying glasses like a glass filled with water. Leonardo da Vinci did not, for all that, look into the question of the two infinites; this went beyond geometry and was a matter for arithmetics (infinitesimal calculus), on which René Descartes would be blocked again more than one century later.

Leonardo could do simple calculations, such as additions. Biographers describe the fate reserved for "illiterate" geniuses, to which he did not fail to compare

1 "One of my treatises on local movement: strength and weight".

himself, conscious of his difference. Since the 14th century the doctors of Florence, in particular, had the tradition of debating through poems in Latin, with an excellent knowledge of the great treatises of medicine (de Libera *et al.* 2019):

> Leonardo, like any man of superior merit, had envious, detractors; he says it himself: as an inventor, one sought to pass him off as ignorant of the works of others. To put it in much the same way as he did, narrow and routine minds on the one hand, and on the other, adventurers on horseback against all that had hitherto made it possible to establish those rules which, determining the limit of the possible and the impossible, prevented the researcher from falling into despair and melancholy, exaggerated with pleasure what he had said in reaction against abuses and paradoxes, and used his own expressions to portray him as a charlatan or a madman. When he spoke at least, his persuasive eloquence gave his ideas their full value; but to allow his papers to show incompletely expressed thoughts, unfinished essays, plans for inventions of all kinds, would have been to expose himself to slander and theft. (Ravaisson-Mollien, *Les manuscrits de Léonard de Vinci. Le manuscrit A de la Bibliothèque de l'Institut*, 1881, author's translation)

2.3.3.2. *Fossils in the Alps and the Po plain at least 200,000 years old!*

Around 1509, Leonardo da Vinci reached the north of the Po plain, Bergamo and then Padua. North of Bergamo, he went to the Alps to study the history of the Earth. He planned to write a book "on the sky and the Earth", presenting his synthesis on astronomy and geology, as indicated by notes found in the Codex F of the Institut de France and that of Leicester: "Knowledge of past times and the position of the earth are ornament and food for the human mind" (Vinci, *Codex de l'Institut de France Manuscrit A*, folio 373 v, 1487–1508, author's translation).

He discovered fossilized shells, folded strata, and observed the powerful work of erosion by the torrents and rivers that ended up depositing sediments in the plain before reaching the Adriatic Sea. These horizontal strata allowed him to approximate the speed of sedimentation and he found an age of 200,000 years. Two sketches of horizontal and folded layers can be seen in the Codex Atlanticus (Vinci, Codex Atlanticus, 1485, p. 493, verso).

> Why are the bones of great fish, and oysters and corals and various other shells and sea-snails found on the high tops of mountains that border the sea, in the same way in which they are found in the depths of the sea. (Vinci, *Codex Windsor*, folio 20 r, 1480–1518)

Inevitably, Leonardo da Vinci doubted the biblical flood:

> Here a doubt arises as to whether the flood in Noah's time was universal or not. It seems that it was not, for the reasons we will explain. (Vinci, *Codex Institut de France, Manuscrit A*, folio 155 r, 1487–1508)
>
> And if you were to say that these shells were created and continue to be created, in such places, by the nature of the spot, and the influence of the heavens, such an opinion cannot exist in a brain capable of much reason (...) how will you explain that this influence produces in the same place shells of various dimensions and different in age and species? And how will you explain that gravel is agglomerated and lies in layers at different altitudes on high mountains? (...) This gravel is nothing more than debris of blunt stones for having rolled for a long time and having been subjected to knocks and falls during the course of the waters that brought them to this point. (Vinci, *Codex Leicester*, folio 80 v, 1508–1510, J.P. Richter's translation, 1883)

It is probable that in the 64 years of Leonardo da Vinci's existence in Italy, he experienced telluric tremors; they were sufficient to establish a relationship between regional-scale vibrations and a continental-scale force that transmits movements over great distances. He did not doubt that the surface of the Earth had changed. Volcanism, although active in Italy and the cause of the death of Pliny the Elder with Vesuvius, was absent, either because these drawings disappeared or because he only drew his own observations. In his notes, he formulated his theory of the history of the planet, that it was a sphere filled with water that caused natural cataclysms and floods, and he imagined it gaining in volume: what was lost vertically was found horizontally and increases its circular thickness. He drew his conception of the planet: it was the first world map with (South) America (Vinci, *Codex Arundel*, folio 236 v, 1480–1518).

2.3.3.3. *An exceptional understanding of the human posture: from verticality to the "gaze of the soul"*

In the first part of his life, Leonardo da Vinci put his genius at the service of the "arts and crafts", but as he discovered philosophers such as Pythagoras and Aristotle, he developed his own philosophy and got closer to the human being. Leonardo was not looking for an architect god with his compass to draw perfect circles and his square with its right angles; his aim was different. It was not a research program but a reflection that evolved with maturity and an iron discipline that he imposed on himself in order to fill the cruel lack of a university culture.

Leonardo received commissions for portraits and scenes from the Gospels. But he preferred to devote the late years of his life to anatomy, to the history of nature and the history of the Earth, and then he placed his characters in real life, which he studied, scrutinized and wanted to understand.

As he filled in his gaps in reading and in encounters with masters, Leonardo da Vinci gained experience and became ever more deeply involved in human anatomy in the search for unity between the osteological framework that struggled against gravity, or Earth's gravitation, the nervous system, physiology and the reflexive consciousness of the human being that passes through the eye and light. He sought the universal creative plasticity of the macrocosm that connected, extended and interacted with the microcosm of organisms. In his view, the human was a living element of a naturally divine reality; the world was a maternal monad. He sought to immortalize the human "soul", the instant of an inexpressible reciprocal emotion, the fleeting moment of a feeling, without any wrinkle darkening it, and he seemed to have succeeded in doing so.

Light, the anatomy of the eye and optics were a recurring theme since it was this close link between nature and his nature that allowed him as a painter to profess his understanding of their links and their principles. It was because he was exceptionally talented in the art of recreating observation and because no academic profession was possible for him that he engaged in this mode of transmitting his own physicalist understanding of nature. His talent enabled him to distinguish outlines, limits, things, reliefs, colors and movements because he thought, because he reflected on the laws of physics. His art became the transmission of an initiation of the reflective regard, not a collection of academic books.

He sought the nature of the link that connected the organization of the human to the world, the being who discovered themself as a sensory "tool", who captured nature through their eyes thanks to their sight and who grasped it through the control of the gesture of the hand that drew it. So what is the seat of this gift that is specific to the human? The brain? The nervous system? From eye to hand, what is the link and who is the guide, what is the nature of this "soul" that wants to understand, that observes and paints the inner reflection (in the nervous system) of reality? His drawings and sketches of the human body and skeleton are studies for paintings that had an almost alchemical purpose in the last years of his life. It was his own laboratory, he explored reality through intuitive experience – the grasping of the line that readjusts itself on the surface of the paper, the sensation that tries to capture volume and movement in a plane – with the perspective and the most accurate rendering possible of what the eyes see, guided by an objective: preserving the integrity of the object observed, while the size of the line increases or decreases according to the distance and position of the observer. He sought to understand the fundamental principles which confer upon humanity the capacity to preserve this integrity of the observed thing whether it is close or distant, the art of the moving surveyor who does not alter the unity of the object, who preserves its harmony.

> The eye, the window to the soul, is the main way in which common sense can simply and magnificently consider the infinite works of nature … If you, historiographers, poets, observers, had not seen with the eye, you could relate nothing in your writings. (Vinci, *Codex Institut de France*, 1487–1508, author's translation)

Leonardo questioned and explored, dissected and drew. He explained visually without comment, because despite his 200 books, progressively acquired thanks to better fortune, he could not read them all; he lacked the vocabulary to understand Latin as well as to write poems. The drawing was worth all the comments, the painting all the emotions and his declamations of grace and beauty.

The most beautiful anatomical plates are kept in the Codex Windsor in the Royal Library at Windsor Castle; they date from the time of Leonardo da Vinci's maturity in the art of dissection (from 1510). Others can be seen in the codices scattered among France, Italy and Spain. These are not works of art that are meant to be exhibited, they are essays and explorations. He began to perform dissections on animals that, like Galen, made chimeras, believing it was enough to modify the size of animal bones to put them back into the human body. Drawings from 1485 to 1490, supposed to illustrate the skeleton of the upper limb of man, correspond to a humerus of a quadruped and the hand to that of an ape (Vinci, *Codex Windsor*, 1485–1490, see online). In the last months of his presence in Florence, he dissected the corpse of an old man and that of a child at the Santa Maria Nuova Hospital. Around 1510, he went to Pavia to complete his dissection with a young professor of medicine, Marcantonio della Torre, who introduced him to Galen's treatise *On the Usefulness of Parts of the Body* and the manual "Anathomia" by Mondino de' Liuzzi. An epidemic of plague broke out the following year in 1511 and took the young 31-year-old professor. Leonardo fled Pavia for Milan, then went to Rome in 1515. The publishing project was abandoned. It was resumed in 1651, 130 years after his death (Fréart 1651).

A Florentine school of surgery was finally established at the end of the 16th century to train doctors, surgeons and pharmacists thanks to a medicinal garden (the Santa Maria Nuova hospital; see Gozzoli (n.d.)). Teaching was essentially practical, courses were given directly by observing the patient or the corpse. This teaching, radically different from that of the University, suited Leonardo, who had no books, did not read Latin and preferred direct observation.

His approach to the most internal territories of the human body, namely the nervous system and its endoskeleton, is the first that has reached us. His drawings preserved in the Codex Windsor are the very first studies of sawn human skulls allowing us to observe the interior of the neurocranium and the face. It is true that Leonardo da Vinci would not have been allowed to dissect at the Santa Maria Nuova hospital if he had not already been taught the art and if he had not been supervised by a "barber-surgeon". However, it is unlikely that Leonardo would have done the skull cleaning and sawing himself; this was the work of a surgeon who knew how to handle the saw and how to proceed with the bone and the dura mater. In other words, the dissecting rooms had osteological collections for the courses and for the surgeons as well. The anthropology collections have skulls sectioned in different planes, so these drawings are not unrealistic, they correspond to the collections of a laboratory and an education in anatomy.

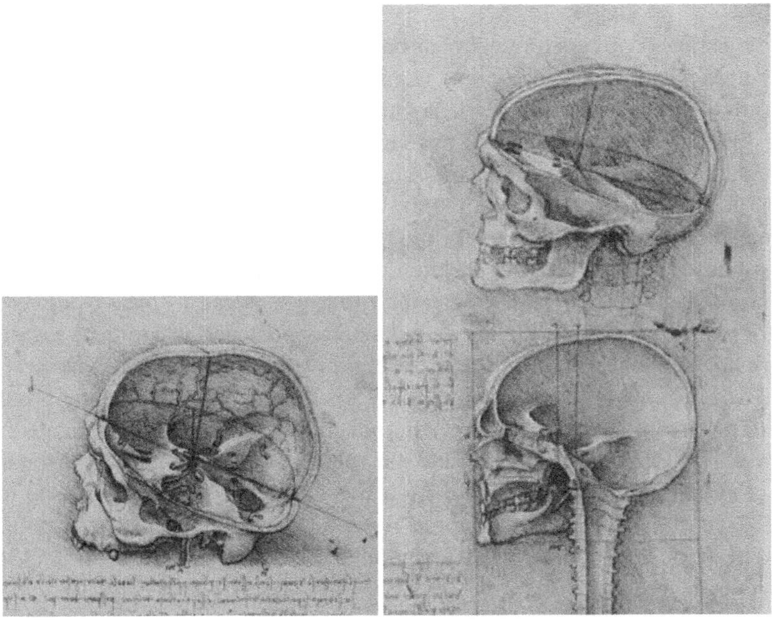

Figure 2.1. *The first geometrical studies of the internal base of the neurocranium and the first study of the relationship between the face and the internal base in medial sagittal section, Leonardo da Vinci, Codex Windsor (ca. 1489). Royal Collection Trust/© Her Majesty Queen Elizabeth II 2020*

Da Vinci depicted a sectioned specimen in the plane of symmetry by carrying out a geometrical study in order to establish linear distance relationships between the exclusively neural, smooth and regular parts with their curves (the vault), and the more complex ones with the two levels of the face, where the organs of sight and phonation, respiration and eating are housed with the teeth (Vinci, *Codex Windsor*, folio K/P 43r ca, 1485–1490). Leonardo also depicted sections of the face showing sinuses (Vinci, *Codex Windsor*, folio K/P 42v, 1485–1490) as well as sections of the neurocranium in perspective with the imprints of the meningeal vascularization on the endocranial surface (Vinci, *Codex Windsor*, folio K/P 42 r or RL 19058, 1485–1490). He sought proportions that illustrated principles or logics of physical, physiological and psychological organization, relationships to establish correlations between structure, verticality and *senso comune* (common sense) or thought.

He noted that the neurocranium is divided into two territories from front to back: the first distinguishes the horizontal zone with the floor of the frontal lobes (including the ethmoid), and the second with the zone that rises up in the center, including the sphenoid and the pituitary gland, and laterally the petrous pyramid (or rock) that reaches the temporal squama bone (which forms the temple). The verticality is accentuated toward the spine with the clivus (part of the sphenoid) and the basioccipital which borders the large orifice (occipital hole).

The two sectors are separated virtually by two vertical axes, perpendicular to the anatomical horizontal. The anterior one passes exactly through the very famous cartilaginous joint, the spheno-occipital synchondrosis (SOS), considered for some decades as the place of flexion of the base. The second axis passes through the anterior point of the large orifice, the place of continuous transition between the spinal cord (vertebral) and the brain stem (cranial), while at the level of the bones, there are spaces, discontinuities. The first vertebra, or atlas, is not in contact with the lower cranial fossa, also known as the cerebellar fossa, and it is articulated with the second vertebra precisely named the "axis", with its excrescence that forms a pivot called the "dens". Da Vinci defined the "fulcrum" at the center of the "dens", at the intersection of the lines r–m and h–f.

The artist understood that the structural and physiological link between the nervous system and the face is located in the center of the base of the skull and concluded that the center of thought is located at the intersection of the a–m and c–b lines (Vinci, *Codex Windsor*, folio K/P 43r, 1485–1490).

We will see, with a geometric reference defined from the embryonic stage (Dambricourt Malassé 1993), how its first vertical axis is close to the axis of rotation which leads to the cordal alignment of the embryo toward verticality. Five hundred years have elapsed between this orthonormal mark and the one drawn to describe the dynamics of embryonic straightening and, therefore, the origin of verticality.

Leonardo da Vinci's drawings gained in quality and excellence. This great explorer of nature and life understood the specificity of human anatomy and how associating with it parts of the skeleton and organs of other animals was an architectural, mechanical, physical and physiological misunderstanding.

Neither a surgeon, nor a physician, nor a chair professor, the finality of his drawings was approached by the spirit of Berengario, who sought to transmit to students and future surgeons a knowledge of physiology. However, da Vinci went far beyond that, because he was above all an artist, an intuitive and contemplative person who questioned human nature with that emotional sensitivity that went beyond the cold and functional calculation of the geometrician.

Leonardo da Vinci situated the center of human thought at the break in the slope between the horizontal territory and the straightened territory of the neurocranium, which does not exist in any other animal, except sketched in apes. Internal verticality is absent in other vertebrates, including apes. Even if he was unaware of the causality, by situating the human "soul" instead of internal verticality, it is probable that Leonardo da Vinci was aware of a relationship between the cognitive and emotional capacities of the human being and this break in slope, because it separated humans and animals geometrically, physically and physiologically. It is also the most internal part of the human body and the most hidden from view. It was necessary to recall that, in 1986, its causality was still not known.

His ultimate quest was the capture of the instant, as if to immortalize it, in what the body and the nervous system possess that is more human than can be transmitted by pictorial art, which is the exchange of glances. He sought the moment of transcendence, of grace that escapes gravity, as Simone Weil would express it five centuries later in *Gravity and Grace* (Weil 1947).

A close look at the portrait of Cecilia Gallerani, painted around 1490, is enough to understand the purpose of anatomical science and the humanistic philosophy of Leonardo da Vinci. This painting expresses the effect of the unexpected appearance of a person in a flow of light. The scene is composed of a young woman and her pet, an ermine lit in the whiteness of its coat. The woman's cornea reflects the light, showing a knowledge of optics, but the appearance is perceived beyond the eye, and the woman expresses an emotion in return. The animal is not despised, it is simply different in its anatomy and the organization of its nervous system; the painter has obviously studied it well because the species is identifiable. The two heads are turned toward the source of light. The human gaze is that of an expressive, vertical, serene and welcoming face, as shown by a slight smile, while the ermine has a long horizontal muzzle, like its body which expresses anxiety; the expression of the gaze is limited to the wrinkled shape of the muscle above the eye sockets, while the immense hand signifies confidence. The woman holds it alive against her heart and a necklace of dark pearls adorn her neck and chest. The inert matter of the mineral can shine in the light and adorn the woman's bust, without harming animal life, while the expression of emotions, eyes, facial muscles and lips underlines the singularity of human physiology. The light thus manifests the appearance of a presence; the woman knows that it is benevolent while the animal does not.

3

The 16th Century: From Generation to Human Physiology

3.1. Ambroise Paré (1510–1590), father of French surgery with "more than barbaric Latin"

Ambroise Paré was born in Mayenne near Laval. He is considered the "father of French surgery". He was not a scholar and his Latin was said to be "more than barbaric". He was a particularly gifted master barber-surgeon who obtained his title of doctor after several years of practice with civilians and soldiers on the battlefields. He was the inventor of many instruments and prostheses. Entering the service of King Henry II in 1554, he tried in vain to heal spear wounds that had perforated Vesalius' face and brain. He devoted the last 15 years of his life to the writing of a treatise in French, *Des Animaux et de l'excellence de l'Homme*, composed of 26 chapters illustrated with numerous figures, which was published from 1575 to 1585 (Paré 1585):

> The status of this work, the first edition of which dates back to 1575, is quite exceptional in the medical production of the sixteenth century, both due to the trajectory of its author and the scientific ambition he manifests. The *Œuvres* of Paré was a true best-seller in surgical literature from the end of the 16th to the end of the 17th century. (Pointe, "Au chevet des Œuvres d'Ambroise Paré", 2020, author's translation)

Paré studied the generation of the human species by describing many cases of malformations (bicephalic, anencephalic, absence of limbs or incompletely developed or supernumerary limbs), as well as cases of twinning, looking for different hypotheses between common sense and popular superstition. He assumed that a child born without a complete leg is a "defect of the seed in two quantities", thus of the father and mother. However, for more serious malformations, he maintained the popular tradition according to which the mother was solely responsible:

> Women at the time of conception and when the child is not yet formed, should not have to watch or imagine monstrous things. (Paré, *Des animaux et de l'excellence de l'Homme*, 1585, author's translation)

The chapter devoted to the central nervous system contains richly illustrated plates which describe precisely the endocranium with the imprints of the meningeal vascularization, the upper part of the brain with the well-detailed convolutions, the cerebellum apart from the brain (p. 189), the brain stem (p. 112), the brain and facial and cranial nerves in profile (p. 114), the whole of the brain from below (p. 133) and finally the spinal cord (p. 116). According to Cuvier, he was the first anatomist to draw comparisons of the bone structure of man with birds:

> He shows that the skeleton of the bird, although it looks very different from the human skeleton, has parts that correspond roughly to those of the human skeleton. It was one starting point for comparative osteology. (Cuvier, *histoire des sciences naturelles*, 1841–1845, author's translation)

In fact, the work contains two skeletal illustrations, one human (p. 571) and one of an ostrich that he himself dissected with a reduced caption (pp. 691–692). He did not make any connection with humans except for the observation "the bone of the tail has nine vertebrae similar to that of Man". Bipedalism was not mentioned. The legs of the bird were hind limbs and not inferior as in humans, its bipedalism resulted from the adaptation of the forelimbs to flight, whereas those of humans were superior. Even a penguin in vertical posture did not have its occipital hole above the pelvis, because its base was horizontal.

> What is more difficult to determine is to what extent he was a true observer of the animal kingdom, that is to say, a zoologist worthy of the name; if he was not above all a compiler of ancient works slavishly repeating the naiveties of Aristotle and Pliny (…). (Germiny, *Ambroise Paré, zoologiste*, 1933, author's translation)

Ambroise Paré reproduced a whole bestiary of legendary animals and monsters which amounted to chimeras between humans and animal species. These associations, which we know do not occur in nature, were not unnatural in the minds of anatomists capable of imagining a woman, though not themselves, giving birth to a malformed fetus due to monstrous thoughts at the time of conception.

3.2. André Vésale (1514–1564), the audacity of objectivity in the face of Galen's anthropo-simian chimeras

André Vésale was born in Brussels in 1514 into a family of Flemish physicians who had been noted for three generations. His father was physician to Maximilian I of Austria (1459–1519), the Emperor of Rome (the Holy Roman Empire), following his grandfather. The latter's father treated the powerful Duke of Burgundy, Charles the Bold (1433–1477). His family thus occupied a certain social position in a Europe bruised by wars between duchies, kingdoms and empires. Vesalius trained at the Paris Medical School by following Sylvius's dissection courses. Tradition has it that Vesalius became familiar with this practice from his youth, accustomed to collecting human skeletal remains from gibbets. Some were visible on the mound in front of the family home. If the tradition is true, he was aware of the anatomo-skeletal differences between the vertical organization of humans and that of quadrupeds. Human dissection was surrounded by taboos and superstitions, but he was not afraid to recover the remains of hanged men and to proceed to describe their organs.

He was called back to Louvain by the emperor Charles V, due to a war against François I, who appointed him first doctor of the armies. Then, in 1536, Vesalius chose to settle in the Republic of Venice, which appointed him professor at the University of Medicine of Padua. He taught there from 1540 to 1544. A contemporary of Copernicus and heir to the spirit of the University of Bologna (close at 160 km), he applied the method of reproducing observations as faithfully as possible. In 1542, he began writing his masterpiece *De humani corporis fabrica* ("On the Functioning of the Human Body") and published the first edition in 1543 with the anatomical descriptions and their illustrations printed in Basel at one of the largest printers in Europe. Vesalius went there to dissect a corpse and reconstruct the skeleton to draw the plates. According to Cuvier, the human skeleton at the Basel Medical School was the first reconstructed anatomical structure in the history of human anatomy. In the same year, Copernicus succeeded in printing his work in Nuremberg just before his death.

Cuvier considered the Treatise of Vesalius as "the starting point of all modern anatomy", "it is from him that we must date the true human anatomy" (Cuvier, *Histoire des sciences naturelles*, 1841–1845, author's translation).

As Leonardo da Vinci never completed the project of printing his anatomical drawings, it fell to André Vésale to produce the first edition of very high-quality anatomical dissections based on his own observations.

His treatise was the very first with descriptions of the bones and excellent figures, as well as a list of characteristics relating to monkeys. The illustrations are said to be by Titian (1488–1576) or even his pupil Jan van Calcar (1499–1545); the frontispiece is perhaps by Veronese (1528–1588). The work is considered "one of the most beautiful books in the world".

The incisor bone, i.e. the bone visible on the palate between the root of the canine tooth and the lateral incisor, does not exist in humans. The human mandible forms a single bone, not two separate hemi-mandibles in the chin area. It also shows the lack of communication between the nose and the brain, more precisely with the pituitary gland, located in the sphenoid, in the center of the base of the skull. Similarly, the two ventricles of the heart do not communicate, Galen's blood circulation model is wrong, as are the two kidneys described in simian and not human position.

The year 1543 thus saw the publication of two major works that were to change the place of the human in the cosmos with Copernicus, and further compel medicine to learn human anatomy and physiology with Vesalius, through dissection and no longer in texts devoid of anatomical plates. The human is not an ape, and knowledge of simian physiology was not enough to cure human physiology. It was thus naturally in Italy that Vesalius taught, not without provoking conflicts based on disagreements of interpretation. The demonstration of Galen's numerous errors was seen as a real sacrilege and led him to be sidelined in the medical community, particularly by Sylvius, Jacques du Bois (1478–1555), who was one of his teachers in Paris for three years. However, not only does the book of Vesalius, better known as *La Fabrica*, question Galen's anatomy and therefore his physiology, but it is also dedicated to Charles V, the enemy of the King of France, Francis I. The affront was to the monarch's status: one was imperial (ruling much of ancient Germania, the Netherlands, Spain, the Kingdom of Naples and Rome) while the other was merely royal.

3.3. Jacobus Sylvius (1478–1555): defending Galen body and soul

Born in Picardy (northern France), from a modest background, Jacques du Bois, or Jacobus Sylvius, owed his fame and social success to the efforts he undertook to return to the original versions of Galen's texts. This work required long years of study to learn Greek and Latin. As such, he is considered by Cuvier as one of the first humanists of the Renaissance (a return to the original texts to fight against medieval scholasticism). Gripped by a passion for Galen's work, he taught him in paid private lessons, even though he was not a graduate. He was widely successful, but since he lacked the title of doctor and the right to teach, Francis I preferred the Italian Guido Guidi, as the first holder, or "reader", of the Chair of Medicine at the brand new *Collège Royal de France*. The *Collège* was founded in 1530 to liberate teaching from scholasticism. Cuvier points out that its reputation was already well established in 1531, but Sylvius did not obtain his diploma until 1535, at the age of 57 years. He then taught at the Parisian college of Triquet (or Tréguier), teaching students to dissect in order to observe for themselves (Goujet 1758). He left a written record of his lectures for the most destitute who could not afford copies of Galen's work, but he refused to reproduce his descriptions in the form of drawings for lack of a mastery of perspective. He made new observations, distinguishing between voluntary and involuntary muscles (which contract without the conscious control of the nervous system); he mentioned the ossicles of the inner ear, whose function in the automatic control of head movements would be demonstrated later. Furthermore, he accurately described the sphenoid bone whose body occupies the center of the base of the skull, as well as parts already known of the brain but which would take his name: the Sylvius fissure, the Sylvius aqueduct at the level of the brain stem – an aqueduct which forms a tube allowing the drainage of cerebrospinal fluid – and the sylvian artery or middle cerebral artery.

> Sylvius, one of the greatest anatomists of his century; he was the first to describe exactly the pterigoid apophyses & the clinoids of the sphenoid bone he admits only three. He gave an exact description of the sphenoidal sinuses of the adult, he knew the palatine bones, the unguis bone did not escape him; he was the first one who gave the name of oblique and transverse to the apophyses of the vertebrae. (Goujet, *Mémoire historique et littéraire sur le Collège royal de France*, 1758, Old French, author's translation)

Now André Vésale, one of his former students, questioned his teaching, which was the masterpiece of his life, by demonstrating that Galen committed important anatomical errors. In addition, too many descriptions referred to animals and not to humans. Galen was still an icon and the popularity of Sylvius, who nevertheless practiced human dissection, was questioned. After the death of Francis I in 1547, Guido Guidi returned to Italy, leaving the Chair of Medicine at the *Collège de France* vacant. Doctors hastened to convince Sylvius to take over. He hesitated for two years. In 1550, King Henry II appointed him to the chair and it was then that he committed an irreparable affront by publishing, in 1551, *Vaesani cuiusdam calumniarum in Hippocratis Galenque rem anatomicam depulsio*: "A Refutation of the Slanders of a Madman against the Anatomy of Hippocrates and Galen". Galen was "the only parent of anatomy", while Vesalius, "this insolent and ignorant", only propagated slander against teachers. Sylvius died on January 13 1555, and his funeral was celebrated with great pomp by the medical corps, not without some mockery[1]. A few months later, a former student, Renatus Henerus, published a

1 "Here lies Sylvius who, during his lifetime, never gave anything for nothing. When he died, he still feels sorry that this line can be read free of charge" (de La Roche & Petit-Radel 1792, p. 544, author's translation).

defense of Vesalius coupled with a systematic attack against Sylvius, *Adversus Iacobi Sylvii depulsionum anatomicarum calumnias, pro Andrea Vesalio apologia*. He described “Sylvian fanaticism” and “the cult of Galen” was all the more “stupid” as, he wrote, Galen was the first to warn the reader that his descriptions were often limited to animals such as monkeys, dogs and pigs.

Irreverence was a style but denial was unworthy. Historians hold it against him for failing to hold his rank as a Royal College professor. Rather than serenely verify the errors, he preferred to defend a thesis fraught with consequences, namely the degeneration of the human body since the time of Galen (de La Roche & Petit-Radel 1792). This theory had been stated by Jacopo Berengario about the ileo-caecal appendage (its inflammation causes appendicitis attacks), but it would have been enough to compare the mummies to demonstrate to the whole of Europe the implausibility of the remarks of a professor of the *Collège Royal* of France. Alexandria traded mummies through its maritime traffic with Venice, Genoa and Barcelona, which were resold in France and Germany, to such an extent that Francis I kept mummies in his personal pharmacopeia as a counterpoison and first aid against war wounds. However, it was above all the insult to four generations of venerable physicians that earned him the blame of history for having deliberately debased, in an article title, their patronymic in the name “Vesianus”, a contraction of Vesalius and *insanus* which means “mad”.

Philip II, the eldest son of Charles V and the King of Spain, appointed Vesalius as his first physician and had him called to his court in 1549. The resentment of the medical profession maintained by Sylvius (Bayle 1697), however, caused him difficulties. In 1561, he narrowly escaped the condemnation of the Inquisition of Madrid. Vesalius had to treat a young woman who died suddenly. He asked for an autopsy. A brother of the deceased had insisted on attending the operation; he reported that the heart had supposedly palpated and testified against Vesalius. As a great specialist in cadavers, it is doubtful that the physician did not observe rigor mortis before operating. The king was undoubtedly not fooled by the intrigue but, faced with the Inquisition, he was forced to go along with it. The sentence was commuted to permanent exile and a pilgrimage to Jerusalem. Vesalius, who was eagerly awaited at the pulpit in Padua, died around October 15 1564 of a bad fever due to poor conditions on the return from Jerusalem. He was buried in Zakynthos, a small island in the Ionian Sea. Historians of science consider Vesalius as a victim, along with Jacopo Berengario, of a false denunciation for not very praiseworthy reasons:

> Sylvius defended Galen with extraordinary liveliness and violence. Everyone knows the crudeness of the literary debates of that time: that of the anatomists was no less. Sylvius' brutality was truly horrible. (Cuvier, *Histoire des sciences naturelles*, 1841–1845, author's translation)

The ape-like characters described by Galen and listed by Vesalius were precise enough to be authenticated three centuries later in Paris by Charles Daremberg, Professor of History at the Faculty of Medicine in Paris in 1871 and 1872; for two years he was able to consult the comparative anatomy collections of the *Muséum national d'Histoire naturelle* and managed to identify the species: it was the *pithekos*, or macaque.

3.4. Gabriele Fallope (1523–1562): freedom of dissection, the fine anatomy of the ear and cranial base

A native of Modena, Gabriele Fallope succeeded the chair of his professor André Vésale in 1544 in a context of rivalry between the Faculties of Medicine of the Kingdom of France and the Holy Roman Empire and Catholic scholasticism. Authorized to dissect human cadavers, he carefully studied the organization of the internal base of the skull and enriched knowledge of the bone structures that make up the ear. His most remarkable observations were printed in Venice in 1561 in the treatise *Observationes anatomicæ*:

> The book was so well received that it was reprinted the following year in Paris. It is full of new observations. The author shows that the skull of the fetus is composed of more pieces than that of the adult. He also shows the differences in the vascular system of the two. The very complicated bone, which has been given the name ethmoid, is better described than in Vesalius. It is also to Fallope that we owe the description of the oval hole of the sphenoid, through which the nerves of the fifth pair pass; that of the sphenoidal and petrous sinuses. He also described the alveoli in which the teeth are embedded; the veins and nerves that go there. What he studied above all was the structure of the inner ear. Fallope discovered the vestibule, the semicircular canals, the cochlea, its spiral blade, the frame and cord of the eardrum, and finally the tortuous canal or aqueduct that still bears his name. He made several important remarks about different muscles, especially the muscles of the ear, both internal or external. I would find it difficult to explain to you this part of his work, which would require details and figures; but those of you who know anatomy know that the external pterygoid is one of the most difficult muscles to observe in man. The muscles of the palate veil have only been well described by Fallope; his description of those of the face is also superior to Vesalius. (Cuvier, *Histoire des sciences naturelles*, 1841–1845, author's translation)

3.5. Bartolomeo Eustachi (Bartholomaeus Eustachius, c. 1523–1562): the human fetus and the monkey

Eustachi was a native of the Kingdom of Naples. He continued to complete human anatomy and physiology. He was particularly interested in the vertical territories of the skull that maintained complex physiological relationships between the ear and the throat (the rhyno-pharynx): he showed, among other things, how aerial communication was established with a duct that took his name, the Eustachian tube. Rivalries with Vesalius were recurrent:

> In all his writings, he defends the ancients against Vesalius with an extraordinary pungency, which almost approaches that of Sylvius. (Cuvier, *Histoire des sciences naturelles*, 1841–1845, author's translation)

Eustachi undertook the first comparative anatomical study of humans and monkeys, from fetus to adult, to ensure that the criticisms against Galen were well-founded. The intention was not to understand the differences but to verify whether Galen was really wrong as soon as Vesalius asserted that a character was that of a monkey.

It is clear that these are not the same growths as he related in 1563 in De dentibus:

> He began the study of organs in the fetus, and continued to examine them at the different ages of the human species. Vesalius had examined the adult, and his work, within these limits, was already quite extensive. But our organs vary with age; almost none of them change in both shape and consistency and proportion at different times of life: and it is obvious that these variations are one of the most important parts of anatomy and physiology to know well. Above all, they are very sensitive for the teeth, since they are not born with us, like our other parts, they only come out successively, some fall out, others succeed them, finally they present continuous variations depending on the age. Eustachi wanted to study these variations; he took them in principle; he began to examine them in the fetus. It was Eustachi who first used this method, which since then has been generally applied. (ibid., author's translation)

In 1564, he repeated the statement in *Ossium examen:*

> which contains a criticism of Vesalius (…). This book is not in very good faith; however, it made the author make observations that enriched science. The result was a good osteology of the monkey, and

> interesting remarks on the osteological varieties of the human species. (ibid., author's translation)

Eustachi left an unpublished work with remarkable anatomical plates dating from 1552:

> Eustachi had prepared another work which would have been more important, if it had appeared in time; these were his anatomical plates, which he had drawn before his eyes to give a complete treatise on anatomy, similar to that of Vesalius, but which would have been much more perfect, since he had added an infinity of objects, and the plates are more careful in detail. They must have been a collection in folio, whose text would have been considerable; he died before he could complete this beautiful work. The brass was engraved in 1552, ten years after the work of Vesalius. They represented many discoveries for their time, but they remained in a store or in some estate for part of the 16th and the whole of the 17th century. It was not until 1714 that they were published by a doctor of the Pope, named Lancisi, with summary explanations. (ibid., author's translation)

These copper plates were announced in the preface to his *Opuscula anatomica* published in 1564 and were intended for a work entitled *De dissensionibus ac controversiis anatomicis* ("Dissensions and Anatomical Controversies"). He was assisted in his work by Pietro Mattei Pini, who inherited all his archives. He justified himself in a letter:

> I am about to publish, one day soon, forty-six copper plates in which this scrupulous application of ours can easily appear, and a book on "Dissensions and Anatomical Controversies". I would have done so much earlier, if I had not been prevented from doing so by my already advanced age, by a very painful pain in the joints, which delays my studies, and by the weakness of my fortune which has prevented such an important work from going to print. But I have not lacked either the will or the desire of my whole soul to do so. (Ruel-Kellermann & Leyrat-Cornuejols, "Introduction à la transcription et à la traduction du Libellus de Dentibus d'Eustache", 2015, author's translation)

These were the 46 plates found among Pini's heirs, 140 years after his death. Pope Clement XI bought them and financed their publication in Rome by Lancisi (1654–1720) in 1714. They will endow him with a late post-mortem recognition:

> They will be republished throughout the 18th century, arousing the admiration of illustrious anatomists including Morgagni, Boerhaave,

> Albinus and von Haller who unanimously found them superior to those of Vesalius. As for the *De dissensionnibus, et controversiis anatomicis*, which were thought to be lost, they were found in the Municipal Library of Siena (Belloni, 1972). Eustachi only continued the polemic started in his Opuscula against Vesalius, without the hoped-for comments to shed light on the *Tabulae anatomicae*. (ibid, author's translation)

3.6. The embryo, the fetus and blood circulation with the maternal body

3.6.1. *Arantius (1530–1589): the development of the human fetus*

Giulio Cesare Aranzio, known as Arantius, a native of Bologna, attended Vesalius's courses and was noted for his work on the development of the fetus, knowledge of which was still very rudimentary, focusing on blood circulation. The study of the placenta, the umbilical cord and a vein that took his name showed the lack of communication between the maternal blood brought to the uterus and the placenta. The placenta is the breathing organ of the fetus. He published his work in 1564 in an anatomical treatise, *De humano foetu libellus.*

3.6.2. *D'Aquapendente (1533–1619): the father of embryology*

Girolamo Fabrizio was born in Aquapendente, central Italy (Lazio). He was appointed to the Fallope chair in 1565. His teaching, which lasted for 50 years, brought the Padua school to the height of its influence. History considers him to be the "father of embryology". His most remarkable works concern the embryology of the hen, a richly illustrated work.

> His treatise on eggs and chicken is very valuable. For the first time, it contains figures representing the development of the chicken from its barely sensitive beginnings to the moment it breaks its shell. (Cuvier, *Histoire des sciences naturelles*, 1841–1845, author's translation)

He published a treatise on the human fetus in its placenta and demonstrated that the envelopes are not always the same between humans and other animals. These were not studies of fetal development proper (Fabrici d'Acquapendente, *De formato foetu*, 1600). He was interested in the blood circulation of the fetus by studying the umbilical cord and the venous canal of Arantius. This vessel carries the blood from the fetal vena cava to the umbilical cord so that its oxygenation takes place in the placenta. Arantius had thus shown that the blood of the mother and the fetus do not communicate.

He also devoted himself to experiments to understand how the voice works. In his treatise on the larynx, he describes how it is produced by blowing into the bronchi. Finally, thanks to "very vast mathematical knowledge", he invented the study of movement according to species, examining running, flying, jumping, swimming and crawling. He also dedicated himself to the comparative anatomy of humans by dissecting several species of vertebrates (reptiles, birds and mammals, including ruminants).

3.6.3. *William Harvey (1578–1657): the demonstration of blood circulation, vital for the development of the embryo*

William Harvey was born in the county of Kent (England). After obtaining an academic education in Cambridge, he attended medical school at the Faculty of Padua from 1600 to 1603, when Aquapendente rediscovered the valves inside the veins.

Back in England, Harvey practiced medicine at St. Bartholomew's Hospital in London from 1609 to 1643. In 1616, he was elected a Fellow of the College of Medicine in London, where he taught anatomy and surgery. Aquapendente died the same year. Harvey was trained in the methods of measurement, calculation and the principle of the reproducibility of observed phenomenon (naturally or experimentally). Motivated by the discovery of valves, he was interested in blood circulation and wondered about the direction of circulation. Pulmonary circulation was known, it was observed between the lungs and the two left and right valves of the heart – but with the erroneous idea of a communication between the valves. On the other hand, it was not in the whole body.

Aquapendente had started the study as early as the intrauterine development, starting with the inferior portal vein through which blood passes from the heart and to the lungs. To verify the extent of circulation, Harvey conducted an experiment on different animals by ligating the vessels and comparing their swelling to the position of the heart. In arteries, the swelling at the ligature is on the side closest to the heart, while in veins, blood accumulates on the side furthest away. The blood in the arteries therefore comes from the heart, while the blood in the veins goes to the heart. He then compared these differences with the direction of the valves, which confirmed his deductions. The blood is propelled through the left valve of the heart into the arteries to the organs it supplies, and it returns through the veins to the right valve. He thus demonstrated the circular structure of the vascular network and the great circulation of the blood. The right valve sends blood from the veins to the lungs and the blood returns to the heart through the left valve, which propels it to the organs. The convergence of venous and arterial bloods in the heart is mostly done without communication between valves. The demonstration, more unified and parsimonious than that of Galen, was published in 1628 under the title *Exercitatio Anatomica de Motu Cordis et Sanguinis in Animalibus* ("Anatomical Exercise on the Movement of the Heart and Blood in Animals"), better known as *De Motu Cordis* "Of the Movement of the Heart").

Galen had noted the difference in quality between venous and arterial blood, and the passage through the lungs signified the physiological role of the latter. The need to introduce air into the body in relation to the circulation of blood was unquestionable. Thanks to Harvey, physiologists were able to deduce that the blood going from the heart to the organs had the right qualities to ensure their functioning, while the blood returning to the heart to be propelled to the lungs no longer did. It did not matter at this point in his immense physiological progress that Harvey was unable to describe how the "black box" of the lung worked. Powerful microscopes would have been needed to see the capillaries. The argument is not enough to justify the denial of Parisian scholastics, such as Jean Riolan and his friend Gui Patin (1601–1672), who were happy to call him "circulator", meaning "itinerant charlatan doctor" (Vons 2016). This rejection was not without worrying consequences since blood-letting was applied in ignorance of this new physiology. Guy de la Brosse (1586–1641), King Louis XIII's ordinary physician and botanist, described these royal physicians as a "bloodthirsty sect". Thus, he won his case in 1636 against Riolan, exiled in England with Catherine de Medici, following the break-up of his chair at the *Collège de France* which compiled anatomy, botany and pharmacy. The chair was split into three chairs for the *Jardin royal des plantes médicinales*, finally independent from the University of Paris in the hands of the scholastics.

> His book initially produced great opposition, as is natural whenever an important discovery appears in a science cultivated by a large number of men and linked to several other sciences. A Scottish doctor, named Primerose, was Harvey's first opponent. Then came Riolan, who was a professor at the Collège de France. These two men wrote at about the same time. Some others, such as Vanderlinden, Hartmann, Almeloveen, etc., wanted to attribute the discovery of circulation to Hippocrates; but all these disputes were only an effect of envy, or rather of the natural inertia of the human mind, which does not like to change and overturn all its knowledge. (Cuvier, *Histoire des sciences naturelles*, 1841–1845, author's translation)

3.7. On human generation and fetal development

Anatomists wondered when the body and skeleton acquired the human form in order to know at what stage the soul could find its matrix. The term "embryology" is broad: it concerns the study of intrauterine development, including the placenta, and therefore does not have the contemporary meaning reduced to embryogenesis (the first eight weeks).

3.7.1. *Gabriel de Zerbis (1455–1505)*

Gabriel de Zerbis was born in Verona (northeast of the Po plain) around 1455 and died in 1505. His university career is not well known. He held the Chair of Medicine at the University of Padua. He became known in particular for his dissections of human fetuses, very rare at the beginning of the 16th century, published in 1502 in the chapter *De generatione embrionis* of his *Opus preclarum anathomie totius corporis humani* ("A Study Shedding Light on the Anatomy of the Entire Human Body"). The work reissued in 1533 was a consequent work but without illustrations.

3.7.2. *Volcher Coiter (1534–1576)*

Volcher Coiter de Groninguen (The Netherlands 1534–1576) followed Fallope's courses in Padua, Eustachi's in Rome, then Rondelet's in Montpellier. He taught at the Faculty of Bologna and was interested in comparative osteology and the embryology of the hen. He gave osteology figures to be engraved on copper to Nuremberg artists renowned for their talent.

As a medical student in Bologna, he obtained skeletons of fetuses and "malcarned" children (malformed), which he compared to those of adults. He published his work in 1572 in his *Externarum and Internarum Principalium Humani Corporis Partium* ("Tables of the Main External and Internal Parts of the Human Body").

He printed the very first, very high quality, illustrations, on two pages, of a complete skeleton of an adult monkey in vertical posture, seen from a three-quarters perspective, and showing the spine, seen from the back.

He added the skeleton of a newborn human baby seen from the front and back, as well as the youngest human fetus compared to the previous skeleton, which gave an impression of gigantism. Most remarkable is the systematic comparison of the differences between the monkey, a *Cynocephalis similis* (Macaca), and humans, with emphasis on the spine, specifically the cervical vertebrae, and the fact that humans have no tail. Coiter illustrated a fine comparison of the internal and external faces of the base of the skull, of the newborn and the adult, in upper and lower views.

On the other hand, there was no comparison of the vertically sawn base in the median sagittal plane, showing the differences in bone straightening between the very young human fetus and the adult macaque monkey. The separation between man and monkey was therefore based on the most visible difference in vertical posture, i.e. the presence or absence of tail. The engravings are visible online (Coiter 1572).

For students wishing to train in medicine and who could not make it to Italy, the key university was Basel, in German-speaking Switzerland, on the border with Germany.

3.7.3. *Félix Platter (Foelix Platerus, 1536–1614), the first optician*

Trained at the University of Montpellier, Felix Platter held the Chair of Medicine at the University of Basel, where he taught Vesalius anatomy, dissected the human body in front of students and published a treatise in 1583, *De corporis humani structura et usu* ("The Structure and Function of the Human Body") (Platter 1583). He had complete skeletal drawings of a fetus and a newborn baby in a vertical posture engraved with precision and finesse, distinguishing the different bones of the base of the skull that were not fused, in endo- and exocranial views. It was the first anatomical plan of the human skeleton at the two-month stage, as shown by the narrow shape of the mandible, which is wide and parabolic at birth.

He added an article on the use of the crystalline lens, "which other anatomists had not dealt with well" (Cuvier 1842–1845), with an exceptional examination of myopia and cataracts. He compared the crystalline lens to convex lenses and the eye to the painter's darkroom. Leonardo da Vinci had described the comparison one century earlier, but it was still unpublished. Nevertheless, Felix Platter's observations were the first to correctly account for the position of the lens with valid schematic illustrations of optics. The article went unnoticed. However, during all these years, while nobody mastered optics, Galileo taught at the University of Padua (1590–1612) and made new astronomical glasses that had the disadvantage of distorting the image. The Republic of Venice was pleased with them because their manufacture allowed the observation of the stars at high magnification for the first time. It is perhaps the fame of Galileo Galilei that prompted Platter's nephew to undertake a second publication in 1626, which was more widely distributed.

3.8. Giovanni Alfonso Borelli (1608–1679): the dynamic geometry of the vertical body

Giovanni Borelli (1608–1679) was born in Naples. He was a mathematician. He founded the *Accademia degli Investigandi* in 1650 in order to stimulate multidisciplinarity and promote the rapprochement of medicine, physiology, mathematics and physics. In 1656, he became the Chair of Mathematics at the University of Pisa and met the anatomist Marcello Malpighi (1628–1694), a native of Bologna. Throughout his life he was involved in the first studies of the static equilibrium of the body and animal movement, including humans. They were published the year following his death (Borelli, *De motu animalium*, 1680).

His approach was geometric and established analogies with architectural knowledge. The basic scheme was the triangle. The notable distances between the points were indicated by letters, their lengths were connected by letters, the whole forming polygons. Since the measurement of an angle did not exist, he took their values to correspond to the opening of a compass between two points and thus the segment of circle that he drew by joining them.

Borelli illustrated this technique by drawing this compass and how to move it, starting from the last point (see figures available online (Borelli 1680)). The basic pattern was repetition of the movement. If it is always repeated in the same direction with the same deviation, the drawing will end up showing a series of zigzags. The vertical is given by the plumb line, which also illustrates the force of the Earth's gravity (Borelli, *De motu animalium*, 1680, see the figures on p. 411).

The limbs were then broken down into bones, joints and muscles. Bones were levers, joints were pulleys with an axis of rotation and muscles were like ropes. Statics was illustrated by the principle of the simplest balance with a beam: an axis of rotation and a weight suspended from each of the two ends of said beam. The weight corresponded to the muscular force necessary to maintain the skeletal framework in static equilibrium. Borelli applied this analogical method to the human mandible for the first time by drawing a weight suspended from a rope, which was placed on the dental arch. This weight illustrated the resistance to the temporal muscle which was well-inserted in the skull and acted in the opposite direction, exerting an upward pull on the rising branch of the mandible which is articulated with the base of the skull.

The movements were therefore geometrized by angle openings and successions of circle segments. The most expressive of the figures was that of a man represented with his muscles in profile and carrying a heavy spherical load on his neck (the atlas vertebra) while holding it with his arms and hands raised. The body needed to use all the muscular forces to maintain balance in a posture, not vertically, but in a succession of triangles that drew zigzags to the tip of the feet. Borelli illustrated the man and woman in movement, either balanced or unbalanced, initiating a fall and indicating the center of gravity toward the center of the sacrum. Finally, all these movements, allowed by the muscles, were driven by the central nervous system. For the doctors it was an "animal spirit" that diffused from the brain through the nerves and reached the muscles. By spirit, we mean a control of movements by the brain via the nerve pathways that act on muscle contractions. Since physiology was rudimentary, anatomists reasoned by analogy. The wetting of the rope modified the tensions of the fibers. It was therefore the blood of the muscles that would be modified in contact with the nerve impulse and which would act on the muscle fibers.

Giovanni Alfonso Borelli was one of the originators of biomechanics, demonstrating the first application of mathematics to the study of the equilibrium of the human body.

4

Centuries in Search of Light

At the beginning of the 17th century, the need was felt to create societies of independent researchers, who did not have to account for their salaries to universities under the authority of theologians. As scientific research was not a paid activity, the researcher had to find a benefactor who shared the same interest in the free exploration of the arts, letters and sciences.

4.1. Independent Academies of Sciences

Francis Bacon (1561–1626) was born in London. He was the author of a treatise on the method of achieving objective knowledge, namely observation, induction and experimentation, the *Novum organum scientiarum*. Induction was not a novelty since it dated back to the first Greek school of Thales and was developed by Aristotle, but it had to be verified through experimentation. For Georges Cuvier:

> Francis Bacon, far from overthrowing Aristotle's philosophy in this way, on the contrary, re-established true physical philosophy as this great man had produced it; he only destroyed the abuse that had been made of his dialectic in the works of scholastic philosophy: so soon this method was universally adopted. (Cuvier, *Histoire des sciences naturelles*, 1841–1845, author's translation)

Naturalists were looking for financial support and a federation of knowledge. This is how the first scientific academy in Europe appeared in 1603, the Academy of Lynx created in Rome by Prince Federico Cesi (1585–1630). He was a scientist and a naturalist, his academy was founded by a pact made with two of his friends. The foundation of a private academy thus marked a turning point in the Renaissance, not only of the arts but also of the sciences, by allowing the purchase of expensive tools such as the newly invented microscope. It also allowed the development of learned societies and networks of competent correspondents. The objective was the publication of the results. This Roman academy disappeared in 1630 with its founder and in a context of great hostility. Indeed, Galileo was one of its most illustrious members, while the seat of the Court of Justice of the Catholic Church, or the Congregation of the Supreme Inquisition, was in Rome. This private foundation was the origin of the present Academy of the Lynx (Italy) and the Pontifical Academy of Sciences (the Vatican).

The Academy of Investigators (*Accademia degli Investiganti*) was founded in Naples in 1650 to support the development of experimental sciences and techniques. Its founder was the mathematician, philosopher, astronomer and physician, Giovanni Alfonso Borelli (Naples 1608–1679).

"The Academy of Experimentation" (*Accademia del Cimento*) was founded in Florence in 1657 by two students from Galilee with the financing of Ferdinand II de' Medici and his brother Leopold. The aim was to verify the statements of physics according to the experimental method. The members carried out numerous experiments in thermometry (heat), barometry (air pressure) and pneumatics (study of gases) through the manufacture of instruments necessary for measurements and their reproduction.

The Royal Society of London for the Improvement of Natural Knowledge was founded in 1660 in London for the promotion of sciences.

L'Académie royale des sciences (The Royal Academy of Sciences of France) was founded six years later, in 1666, by Jean-Baptiste Colbert (1619–1683), minister of Louis XIV, and occupied the King's library in the Louvre Palace, a surface area of 500 m^2. Its vocation was the development of knowledge but it was also political because its members needed to advise the king within the framework of their competences. Chosen by Colbert, the new members were then appointed by the king on the Academy's proposal. According to Cuvier, it became, by its influence, the premier academy of sciences in Europe during the 17th and 18th centuries.

Mathematics, along with calculation and operations, allowed the development of the physical sciences: kinetics (measurement of speed) and kinematics (measurement of motion). Motion was studied as a function of time and space, regardless of its still problematic cause. Time was therefore a parameter in the measurement of speed and motion, but it was not yet a duration as a consubstantial physical property of the thing "in itself". The cause of displacement or movement, the *impetus,* had not yet been formulated. It would be force, which will not be formalized before the recognition of universal gravitation by Isaac Newton (1642–1727).

4.1.1. *Gerolamo Cardano: of the necessity and the form of Man, by spontaneous generation or by putrefaction?*

Cardano (1501–1576) was born in Pavia. His father, a friend of Leonardo da Vinci, gave him a solid education. He became a doctor and also a mathematician, philosopher, astrologer and therefore a mage. From 1550 onward, Cardano began publishing a sort of encyclopedia entitled *De Subtilitate* (Cardano 1556) in which he took up Aristotle's gradation with a clear influence from Leonardo da Vinci (Bourdier 1960).

Aristotle's supra-lunar world (being, soul, God) and the sublunar (physical) world formed a single reality, the first movement remaining in the second movements, the world becoming a divine immanence ceaselessly creative; "nature is a continual creator of living forms" (ibid.), a "huge living body" (Margolin 1998).

The ninth book is titled "Of animals generated by putrefaction", the tenth book "Perfect Beasts" and the eleventh book "Of the necessity and form of Man".

In the ninth book, the Italian physician thus takes up the question of the generation of animals that are born from putrid matter and are therefore not "perfect" animals (Cardano 1556, pp. 195–198):

> (…) why do ants, butterflies and locusts (grasshoppers) produce eggs (…)? It is necessary to believe as many small beasts, and also larger ones to be engendered of putrefaction, given that it is already evident in mice (…). As I have said, when putrefaction precedes, the fat humor is separated from the ashes, the heat that comes on the spot ("incontinent" in the text) places within it the soul fit for such matter (…) from which it has come that all beasts that are engendered from putrid matter, like the worm called "millipede", finish the delivery pains since they are formed of very hot and very wet matter. (Cardano, *De la subtilité et subtiles inventions, ensemble les causes occultes, et raisons d'icelles*, 1556, Old French, author's translation)

The soul remains the principle that gives movement to the form and organizes it. The mage-doctor rather drew up a sort of inventory of the physical conditions of the substrate that allowed the soul of a given species to find the context favorable to its materialization. No novelty can be added to Aristotle's works; these writings are not the result of personal observation, but of repetition of the master's writings.

> It remains that we show why (some) animals are generated from putrefaction, the others by propagation and lineage (…) the cause of this difference is that Nature has wanted this failure to happen to few

> of them: for this cause mainly it would have sought putrid matter to all animals, but for those perfect ones that would require long time to be accomplished, matter cannot be preserved for so long a time without motion and mainly without conceptacle because of the change of time, for these causes the matrix has been necessary. (Cardano, *De la subtilité et subtiles inventions, ensemble les causes occultes, et raisons d'icelles*, ninth book, 1556, Old French, author's translation)

In other words, for the oviparous and viviparous, far less numerous than the insects, the movement of construction of the organism is much too long, in comparison with a fly, to be carried out from putrid matter, as for certain insects:

> The soul, therefore, is determined by celestial heat: and all species have their number. And the species is not moved by matter, if this one cannot receive form (…). The heat that maintains and nourishes the hen's seed converts the white into a fistula (…). And when the heat is further increased, the spirit almost pierces the matter that is the egg or the hen, until it has reached the membrane of the egg. In the beast which has four feet, the spirit comes to the womb. Warmth, therefore, movement, the fistula of the spirit, maintained until the animal is perfect. (Cardano, *De la subtilité et subtiles inventions, ensemble les causes occultes, et raisons d'icelles*, eleventh book, 1556, Old French, author's translation)

In this Aristotelian (the soul or *anima*), Pythagorean (number) and magical (the celestial heat allows the elementary heat that emanates from fermentation) syncretism, let us understand this: the species only takes shape if the soul that corresponds to it finds the potting soil (the decomposed matter) or the matrix (the egg, the uterus) at the measure of its degree of perfection (of completion) and can unfold there.

It therefore took a very long time to perfect the viviparous, which was conceivable for Cardano since he had been aware of Leonardo da Vinci's work demonstrating the beginning of the formation of the Po plain 200,000 years ago. Since long durations were conceivable, he stated the possibility for a species to modify itself according to the circumstances of the environment by taking the example of the wolf and the dog.

> Dogs are the most perfect of this kind, which in the past have originated from the wolf. When they are bred they first turn into wild dogs and then into wolves, just as tamed wolves finally turn into dogs. (Cardano, *De la subtilité et subtiles inventions, ensemble les causes occultes, et raisons d'icelles*, ninth book, 1556, author's translation)

It is the example of domestication which reveals a certain plasticity but always within the limits of the original form; since this plasticity is reversible, it does not change "essentially". In reality, it is a question of the variety of a species and not of the transformation of the species, the wolf and the dog are two subspecies of the *lupus* species of the genus *Canis*.

He took up the problem posed by Aristotle of the possibility of two species reproducing with one another, using the example of the mule born from the male semen of the donkey and the female "seed" of the mare. Two "forms", or "animal souls", can give birth to a new "form". If the two forms donkey and horse harmonize, Nature is satisfied, but the new variety is sterile. In conclusion, Mother Nature never designed the concept of the mule and, left to their own devices, "her creatures do not create new species". Cardano can therefore in no way be considered as a founder of transformism.

Humans are treated in the eleventh book, *De la nécessité et de la forme de l'homme*, which begins thus: "Why Man was made". Three centuries have passed since Thomas Aquinas. The sensitive soul, that of the senses or instinct, is peculiar to animals, and the intellectual soul, or reason, is peculiar to Man. Cardano starts from the biblical principle according to which the human being is a cosmic finality and Man is like an animal, but he is apart by virtue of his soul which is not animal.

The mage approached the question of the generation of Man through fetal malformations:

> And when monsters (fetus) move away from the human form, so much the more when approaching brutes, wild beasts, fish and snakes, so many less survive (…). Whether a monster is a simple error of Nature or whether Nature claims to some finality, such as making a ram out of a man, it must be shown. For if Nature had no finality what would be born would be formless, but that is not always formless. None has perceived as much as Aristotle the limits of power of Nature to go beyond the gender and the genus when it wanders: it is a female and not a male that begets sometimes fish but never trees (…). It seems that when Nature cannot achieve its purpose, it makes genus an end in itself (…). For this reason we constantly doubt whether monsters have anything to foretell of a new genus. This belongs to the books of the Secrets of Eternity. (Cardano, *De la subtilité et subtiles inventions, ensemble les causes occultes, et raisons d'icelles*, eleventh book, 1556, Old French, author's translation)

His reasoning shows a certain willingness to rationalize the observations. Fertilization between the two sexes of the same species does not produce a disorganized or shapeless entity. There exists, in short, a determinism between the animal-physics, the seed, and the eidos, or the animal-mind, which can be deployed there. These observations as a physician gave him the opportunity to see the different aborted stages of the intrauterine development of Man. What does he observe? He had reason to ask himself some good questions, because what he saw is reminiscent of the gradation of the natural history of animals. Now, even if species seem to be sliding imperceptibly toward Man, none is the purpose of the one that precedes it in ascending order. So what is the form of Man and his soul, if his development is a succession of forms that passes through the fish stage and then the snake stage, or that ceases at the fish stage, or at the snake stage, by aborting? Or that becomes totally monstrous? What is this soul that can engender a being that is not viable and resembles nothing already known? Is it nature that continues its creation with errors? The paradox is destabilizing between the soul that must be divinely perfect and a physical failure. Are these monsters omens? Or are they not the wrath of God against unnatural morals or bad thoughts of the mother during mating?

> Monstrous touchings are given to them, absurd imaginations, and pravity (depravity, perversity) of temperament. And for such childbirths that are not right, they are referred to the ire of God. (ibid., author's translation)

Since the soul was thought to be of divine essence and the female body was not thought to be divine since Aristotle, Cardano concluded that the malformation came from the female matter. Cardano was a pure product of the 16th century, politically troubled and confused in his interpretation of nature, one that mixed magic, sorcery, number and alchemy. Irrationality and superstition gave free rein to a detestation of women. Renaissance studies have amply demonstrated that this century was the most murderous for midwives and healers, as well as for mothers giving birth to "imperfect" children, condemned to torture and death for trading with the devil. Rene Descartes reached the opposite conclusion, God alone is at fault for malformations.

Concerning the moment of procreation or the generation of the human body, the mage-physician did not say anything particular. The human body was thought to be engendered by the union of two other human bodies. The generation took place in the womb and it was the spirit of its species that was incarnated there. Some authors attributed to it the following assertion: "the rotten organic matter that was an ape can give birth to a man by virtue of the shared properties of these two species". However, there is nothing in Cardano's work that refers to apes and that allows us to link to this mage the first sign of an intellectual libertinage – one that saw the decomposed body of a cynocephalus as a uterus to receive the male and female seed of Man.

On the other hand, indeed, Cardano wondered if a mouse formed from degraded matter would be the same species as a mouse born from a mouse, or if it would not be rather a monstrosity resulting from the matter and not from the warmth of the species' soul:

> Are mice engendered from putrid matter of another species than those engendered by succession? Certainly not, for the form and operations are similar, but only they differ as the monster differs from the whole beast which is of its species. Thus we will say that the mouse born of putrid matter is a stupid and monstrous mouse. I am amazed by Aristotle for the fact that in this cause he preferred matter to form. (Cardano, *De la subtilité et subtiles inventions, ensemble les causes occultes, et raisons d'icelles*, 1556, author's translation)

Nor does there appear any trace of this observation in Aristotle's "The Generation", so it could well be a legend copied for centuries by an error in translation from Greek to Latin. To give birth to mice that have an umbilical cord, well-known to the Greek naturalist in old clothes as a womb, is to insult him, because the cord was already recognized as the part that nourishes the embryo from the mother's body. Cardano was under the influence of the famous alchemist Paracelsus (1493–1541), whom he visited and for whom putrefaction was a transmutation: "It consumes bodies and changes them into new substances; it produces new fruit. Everything that is alive dies, and everything that dies is reborn" (Paracelsus, *Œuvres*, 1589–1595, author's translation).

4.1.2. *Giulio Cesare Vanini (1585–1619), "Prince of the libertines"*

The "Prince of Italian Libertines of the 17th century" was a defrocked and rebellious Carmelite monk. Originally from the south of Italy, trained at the school of Averroes of Padua, his critical thought was built up during his peregrinations toward the north of Europe, confronting the different philosophical schools with Roman (papal) scholasticism, enjoying the full force of its Supreme Inquisition (Foucault 1999). Vanini wrote a dozen works, including one written in the form of a dialog in which he masqueraded as a disciple of Cardano, exposing himself to censorship and the death sentence for heresy (Smith 2007). Julius Caesar Vanini was mocking (Charbonnel 1919), he appeared more as a provocateur than an intellectual libertine, distancing himself from scholasticism and observing nature by himself.

He was accused of atheism and, at first, in order not to upset his peers, he made a first attempt: "*Amphitheatrum aeternae providentiae, divino-magicum, christiano-physicum, necnon astrologo-catholicum*, against philosophers, atheists, epicureans, peripateticians and stoics".

However, he persisted in his approach, producing a work in the style that would later make Molière successful, a work in which he put his character Julius Caesar in dialog with Alexander, who considers himself educated but who is naive in reality, taking what he has learned from the scholar literally. One like the other is a caricature of a society that is infused with science and philosophy. In the chapter "On the Origin of Man", Vanini mocks:

> Others have dreamed that the first man was born from the rotting corpse of monkeys, pigs and frogs, because between the flesh and morals of these animals and man there is a great resemblance. (Vanini, *De l'origine de l'Homme*, 1842, author's translation)

It is a satire, a parody, Vanini makes a mockery of the literal interpretation of Genesis 1, the divine breath that transforms dust into a human couple. The same breath was mistaken in drawing lecherous humans from the putrid flesh of apes. Above all, he points to the duplicity of a society that claims to follow religious precepts. His freedom of tone earned him the description "convicted of blasphemy, impiety, atheism, witchcraft and corruption of morals" at the Parliament of Toulouse. He was condemned to torture on February 19 1619, had his tongue cut out and was then strangled. Resisting to the end, as if to signify his freedom of speech, he opposed the executioner who finally cut out his tongue. His body was burned and his ashes scattered.

As for Gerolamo Cardano, Julius Caesar Vanini cannot be credited with the first transgressive thought granting an animal and simian origin to the human anatomy. This would be an anachronism; his mockery concerned the matter in which the divine breath would have incarnated, undoubtedly a pig or monkey corpse, in light of their morals. If Vanini had wanted to propose a rational model to make the human anatomy a transformed ape, he would have begun his reasoning starting with the ape's womb. He did not, which would have been the case if his intention had been to question the most rational separate creation of species and classes. Likewise, it would be a betrayal of the meaning of his satire to make him a martyr condemned to the stake by the Catholic Church, for it was not the Inquisition that pronounced the sentence by depriving him of a Christian burial, it was a certain civil society in Toulouse that he mocked in reference to Machiavelli (Cavaillé 2001).

4.1.3. *Man absent from himself, God always as explanation*

A few literate women, like Anne Conway (1631–1679), had the "privilege" of participating in the philosophical debates of the late 17th century. Intellectuals were more interested in being or God than in human beings and their cognitive capacities. It was an explanation of the world that was sought, not an understanding of the human singularity with the ability to seek the explanation and meaning of the human existence.

For these learned philosophers of the 17th century, among them René Descartes, tending toward the indefinite smallness of each of the atoms of the body of Man did not lead this rationale into the void and thus into an interior absence. It was not a regression toward nothingness, it was on the contrary intended to tend toward its origins, or the plenitude of the being that the individual soul finds at the end of its singular generation by abandoning its incarnation at the moment of death.

Blaise Pascal expressed well this awareness of knowing oneself to be between two infinities, this vertiginous feeling meditated during philosophy classes:

> For after all what is man in nature? A nothing in regard to the infinite, a whole in regard to nothing, a mean between nothing and the whole; infinitely removed from understanding either extreme. The end of things and their beginnings are invincibly hidden from him in impenetrable secrecy, he is equally incapable of seeing the nothing whence he was taken, and the infinite in which he is engulfed. (Pascal, *Pensées*, 1897, Paul Kegan translation, 1901)

4.2. The beginning of Man and Russian dolls

4.2.1. *From microscope to microcosm*

4.2.1.1. *Van Leeuwenhoek: a Dutch draper who discovered the microcosm*

Van Leeuwenhoek (1632–1723) was a draper in Delft (Holland). He would have started making magnifying glasses, probably as a linen tester, and thus would have made lenses for a Flemish naturalist, Jan Swammerdam (1637–1680), who devoted his time to the microscopic study of insects. Swammerdam was established in Amsterdam and he soon went to live with his father, an apothecarist who had a cabinet of natural curiosities brought back by the ships of the Dutch East Indies Company. He was an annuitant who sought isolation. Georges Cuvier mentioned him during a lesson by evoking "the system of evolution" involved in the nesting of forms during fetal development and insisted on the importance of the concepts developed in *L'Histoire générale des Insectes* (Van Leeuwenhoek 1669).

Swammerdam discovered metamorphosis, but for him it was the continuous development of a seed that contained all the other seeds of creation until the end of time. This is the dogma of nesting, like the Russian or matriochka dolls.

During this time, three compatriots, Johannn Van Horne (1621–1670) in 1668, Theodor Kerckring (1640–1693) in 1671 and the physician Régnier de Graaf (1641–1673) in 1672, developed the doctrine of ovism according to which the uterus contains a true ovary, or "egg-producing" factory.

De Graaf described his procedure and observations in his "Anatomical History of the Genitals of Male and Female" after dissecting a group of rabbits which he followed from mating to birth. He concluded that the embryo is in the egg before the arrival of the male semen. The latter is limited to giving life to the embryo through the remote action of its *aura seminalis*.

Antoni Van Leeuwenhoek adhered to this doctrine. He had free time and devoted it to the study of microscopic organisms. He improved the optics to magnify the observation scale by 200 times. The quality of the lenses was exceptional and this is how he was able to describe protozoa, bacteria and his own spermatozoa for the first time. After his articles addressed to the Royal Society in London it took a few years to obtain the high performance lenses needed to verify his descriptions. They earned him his election to the Society in 1680, then to the *Académie royale des sciences de Paris* in 1699. The organisms, which he called animalcules, reached sizes smaller than a millimeter; these microscopic scales still remained animated, organized and capable of reproducing. Van Leeuwenhoek was therefore interested in reproduction and the problem of the spontaneous generation of metaphysical essence. Microorganisms seemed to be formed from rainwater, but he later claimed that the drops actually contained seeds. Thus, even the tiniest organisms were born by fertilization of seeds deposited in a physical matrix, water or putrid matter, and not preformed directly from water, like an adult mouse from old dirty matter.

Spontaneous generation was no longer sustainable. He learned of the famous experiment by the Italian Francesco Redi (1626–1697), which invalidated the dogma of spontaneous generation (Redi 1668). This experiment consisted of placing two pieces of the same meat in a large jar. The first was in a small jar covered with a cloth that allowed the air to pass through, the whole located in a large jar where the second piece was deposited. The large jar remained open. Maggots did not appear in the small jar; he concluded that their eggs were laid by flies on the meat. Francesco Redi was also the founder of parasitology and he described more than a 100 species of parasites.

In 1683, Van Leeuwenhoek contradicted ovism and asserted that the human being was potentially in the animalcule and not in the ovaries:

> It is exclusively the male semen that forms the embryo, and the only contribution the woman can make is to receive the semen and nourish it. (Fischer 2003, author's translation)

The two doctrines were thus opposed between animalculism and ovism. It was then that Leibniz intervened in search of a philosophical system capable of integrating metaphysics and physics without confusing them, metaphysics being necessary to understand the origin of the physical laws that govern the development of forms and thus organisms.

4.2.1.2. *Leibniz: the potential embryo-producing cosmos*

Gottfried Wilhelm Leibniz (1646–1716) was born in Leibzig, Northern Germany. He received a multidisciplinary education, first in philosophy and then specializing in law, which he would later make his profession. He also learned mathematics from a Pythagorean, Erhard Weigel (1624–1699), and made mathematics his field of choice. His system of thought represented the synthesis of the scientific knowledge of his contemporaries and involved more than 20,000 epistolary exchanges.

Leibniz was interested in infinitesimal calculus to tackle the problem of the infinitely small and thus the problem of the beginning. Since Newton, naturalists could not go back to the beginning any further than the force of attraction which was universal and present at all scales of the real or physical world. The force of attraction attracts, or separates, the smallest entities and generates their movements. The origin of attraction is metaphysical. The origin of the beginning of an organism remains metaphysical and this beginning is a set of attractions that will form the organism; this set of dynamics is specific to each species or genus.

Leibniz abandoned spontaneous generation and reconciled ovism and animalculism in a doctrine of *"ovo-vermiste"*. The dynamic model of the embryo pre-existed in the egg and in the animalcule (animalculism). The two seeds are therefore necessary for the dynamics to actualize and generate the organism and the stages of its construction.

This was a first point. The second point was brought by the discovery of microorganisms reproducing in the bodies of "infinitely" larger animals (parasites). Leibniz gave himself a representation of the relations between the infinitely small and the infinitely large (the cosmos), which allowed him to approach the generation of species in a very original way. He generalized this organization of smaller "animals" within larger animals to the whole universe. The world became a universal monad that contained different levels of realities subordinated to each other and nested within each other, monads within monads. This monad was a great living organism in which the human was a monad but one with a singular place.

Indeed, for Leibniz, if human beings define themselves as intelligent beings capable of hierarchizing the world that contains them, by their metaphysical thought and by the deciphering of their physical laws, it is because they are the inner mirror of the world, their conscious reflection. Thus, the human being can reach the ultimate explanation, their "raison d'être"*, in this universal monad with which they have a consubstantial relationship: to discover themself as an emanation for contemplation.*

The human being is thus a metaphysical being whose incarnation is potential in the dynamics of their species (or its morphogenesis), contained in the animalcules and eggs. The dynamics complement each other and the metaphysical being unfolds through their reciprocal fecundation.

This was Leibniz's "theory of development" or "genetic law".

The animal is therefore not an automaton, i.e. inert materials moved by external forces, as Descartes had conceived it. The forces are at the beginning, they do not create the being, they deploy it in its stages of construction, they give it an interior animation. A fertilized egg is a living being with its own potentialities, inherited from the egg of the mother and the animalcule of the father.

Leibniz had knowledge of the fossil bones of rhinoceroses, elephants and hippopotami collected in Europe and he concluded that the climate had changed; it was wetter and warmer than in the 17th century.

He was also aware of the fossilized shells that forced philosophers to admit there were geomorphological changes of the Earth's surface on a global scale. These "petrified animals" unknown in the era were well acknowledged; they were said to be "lost". Leibniz therefore integrated long geological timespans and questioned these geological and geomorphological events.

A text entitled *Protogaea, de l'aspect primitif de la Terre et des traces d'une Histoire très ancienne que renferment les monuments mêmes de la Nature* was published after his death (1714) (Gohau 1994).

The series of fossil species continued up to his own day, prompting him to imagine lineages that go back in time and are scattered over the surface of the Earth according to climatic fluctuations. The German philosopher made parallel lineages, *Aristotle's classification* became *a slice of present time*, the animal and plant worlds – both those disappeared and those still living – became a kind of virtual, vanished "great forest", composed of countless species, each forming a "trunk" that sinks into the past to be recomposed.

In other words, the dynamics of each primitive monad that engenders the human, the cynocephalus, the elephant, the snake, the fly, the oyster, the protozoan, are in their seeds and they go back to the creation of the world, they are specific monads. The variation of species can be seen over time, depending on the environmental circumstances, but the variations remain within the limits of monads well separated from each other. The discontinuity between classes or grades is radical; each monad goes back to the creation of the world and originates without discontinuity in a metaphysical reality.

However, the dilemma of the chicken and the egg remained, because it was impossible to see how human seeds would emerge on the Earth in a puddle of water before developing into man and woman and without an umbilical cord at birth.

Human thought was not yet mature enough to understand that the first embryo of a "primitive" human being formed with its placenta in the body of a simian mother.

Epistemologists consider that Leibniz would not have ruled out evolution when he wrote:

> Is it not conceivable that in the great changes that the globe has undergone, a large number of animal forms have been modified? (*credibile est per magnas illas conversiones etiam animalium species plurimum immutatas?*). (Leibniz, in *Gohau Analyse d'ouvrage G.W. Leibniz* "Protogaea – de l'aspect primitif de la terre", 1994, author's translation)

Leibniz's theory of development is not, however, the theory of evolution from protozoan organization (unicellular), for which, in fact, the question of its formation arises from elements that tend toward the infinitely small, with earth (clays?), puddles and very specific attractions, because they will have the property of reproducing themselves.

4.2.2. *The created species are not immortal*

In order to admit a modification of species, one needed to first accept that they could cease to exist in order to modify themselves, inconceivable in view of the dogma of the immutable and perfect creation of species; nor is it conceivable without genuine paleontology collections as the basis of a vast, long-term research program. We had to wait for Georges Cuvier.

Leibniz is famous for his formula “nature does not take a leap” (*Natura non facit saltus*). Nature being one and indivisible, it is a metaphysical reality composed of physical discontinuities.

In the physical world, objects, animated or not, are bordered, limited in their space by a surface. In the same way, species are generated independently of each other, no continuity unites them in the depths of time. By “nature does not take a leap”, it should be understood that monads do not physically pass from one class to another, even though long geological timespans might allow it.

In the end, the mathematical philosopher designed mathematical species that could generate geometric forms just by dynamizing them to give them a bordered physical reality.

Physical species were defined by the possibility of returning to the original form through generation, but no seed could generate a new mathematics with its associated dynamic principles; this potentiality of renewal remains that of being, it is the “mens momentanea” *or the “momentary spirit”.*

Only extinction is physically possible, as soon as living conditions no longer allow reproduction; but the appearance of new models, or types, cannot be physical, it remains metaphysical.

A handful of naturalists did not “refuse the scandal” of the mortality of species (Balan 1979) at the risk of confronting the dogma of an immortal creation. In 1668, Robert Hooke, physicist and member of the Royal Society of London, read articles on geology to his colleagues:

> He explained that fossils were the remains of living organisms and that forms that existed only in warm seas had once lived in Europe, while other fossils differed from all known species. He deduced that sedimentary rocks had recorded the great contemporary changes in the *disappearance of certain groups of animals.* He then described earthquakes and, observing that they caused rock upheavals, he saw one of the causes of orogenesis. He also pointed out that fossils were used to classify rocks because the deepest fossils marked the oldest rocks, even when they had been uplifted. He finally thought that fossils would one day make it possible to date the past. (Birambaut, *Fontenelle et la géologie*, 1957)

Hooke attacked dogma by showing that there was no proof that there was no earthly end to the generation of species. For the idea of evolution to be conceivable, it was necessary to accept the "mortality" of species, the "vulnerability" of a level of reality dependent on time and space.

In 1706, Bernard Le Bouyer de Fontenelle (1657–1757), then perpetual secretary of the *Académie royale des sciences*, returned to Leibniz's observations in the schists of central Germany:

> Mr. Leibnitz conceived that some kind of earth covered the lakes and meadows, and buried fish and plants in them, or that some muddy water laden with earth enveloped or washed them away. This earth has since hardened into Slate ... there are several indications that there must have been great physical changes on the surface of the Earth. Mr. Leibnitz believed that the Sea once covered almost everything, and that afterwards a large part of its waters made a passage to enter into hollow abysses, which are inside our Globe. From there came the Mountain Shells. (Birembaut, *Fontenelle et la géologie*, 1957, author's translation)

For the time being, it is the Earth that has a disappeared history: the cartography of the land and the emerged islands has changed; this is already a considerable awareness. The relationship between climate change and the conditions of existence of species is not an immediate subject of reflection; paleontology, or comparative anatomy, and the classification of fossil species are still in the limbo of knowledge.

5

The Century of Naturalistic Enlightenment

5.1. The *Jardin royal des plantes*: a new natural history of animals

5.1.1. *Georges Leclerc, Count of Buffon*

Georges-Louis Leclerc (or Leclerc) (1707–1788), Count of Buffon, was born in Burgundy. His personality and his work were typical of a researcher trained at the *Muséum national d'Histoire naturelle de Paris*. A monumental statue faces the *Grande Galerie de l'Evolution* in the *Jardin des plantes* (Royal Botanical Garden), while that of Jean-Baptiste Lamarck (1744–1829) welcomes visitors at the other end of the garden, facing the Austerlitz Bridge which crosses the Seine. His naturalist and scientific stature make him a historical figure of the great period of transition toward transformism. Georges Leclerc spent his childhood in Montbard in Burgundy, then engaged in multidisciplinary studies, in law, medicine, botany and mathematics; he read Newton (still alive at the time) and was passionate about his study of infinitesimal calculus. Travels in Italy and England enabled him to learn English, so he began to work on various translations, including a study of Newton. He inherited a forest estate in the Montbard region named Buffon, which he added to his family name. He maintained and operated it with a good knowledge of the mechanical properties of wood. In 1731, the *Académie royale des sciences* was charged by Count de Maurepas, Secretary of State for the Navy, to develop research on these properties. The Count wanted to be able to select the best wood for the kingdom's naval constructions. Leclerc settled in Paris in 1732 at the home of the king's apothecary, Gilles-François Boulduc, member of the *Académie des sciences* and demonstrator (or lecturer) in the chemistry chair of the *Jardin royal des plantes*. Since the Academy could not find anyone qualified in its ranks, Leclerc was solicited and willingly lent himself to tests and comparisons with his own forest species.

His reports were appreciated and introduced him to the court of Louis XV, which allowed him to frequent the circle of Parisian academicians. There he discovered gambling, in particular *franc-carreau*, which consisted of throwing an ecu (a round-shaped piece) on a tiled floor and betting on where it would fall: the bet was *franc* if the ecu fell onto a square, it was lost if the circle cut at least one straight line, the maximum being four straight lines (intersection of four squares). In 1733, he invented a method of calculation that took into account the shape of the object thrown and found that the results were different between the square and circular shape. Comfortable with Newton's infinitesimal calculus, he took up the problem of squaring the circle[1] and submitted his manuscript to two members of the *Académie des sciences*. One was the mathematician Pierre de Maupertuis (1698–1759), a fervent defender of Isaac Newton and universal gravitation. The second was a young and brilliant mathematician, Alexis Clairaut (1713–1765), who had just been elected to the Academy in 1731 at the age of 18 years, before the legal age of 20 years. He had a passion for games that curved forms between them, hyperboles, ellipsoids and the true shape of the Earth that had lost its curves, torn between the elongated shell and the flattened sphere.

Leclerc was 26 years old, Maupertuis 35 years old and Clairaut 20 years old. All appreciated the new method and had it published in the *Mémoires de l'Académie des Sciences*. Indeed, it was the first time that infinitesimal calculus, which poses the mathematical problem of squaring the circle, was introduced in the calculation of probabilities. It is a problem of shape at the limits, and not of physics with too many parameters (force of throw, therefore quantity of movement, height of the arm, resistance of the surface to air pressure, good visual optics, etc.). This innovation earned him his election as a member the following year, at the age of 26, as assistant in the mechanics class.

Randomness (the unpredictable) in natural phenomena, such as the trajectory of an object, is not a new idea; the Greek philosophers, and Aristotle in particular, already reasoned about the notion of chance. This was not without interest for study of the natural history of animals and the classification of new varieties. In 1734, Alphonse de Vignole (1649–1744), president of the Berlin Academy of Sciences (founded by Leibniz), took Genesis from the Bible literally and deduced that the world was no more than 9,000 years old, whereas Leonardo da Vinci had already calculated an age of 200,000 years for the sedimentation rate in the Po plain. Scholasticism remained locked in its bookish rigidity.

1 Calculating the area of a circle with squares posed the problem at the boundaries. Physical contact is a point, but in geometry, the shape of the "contact" was pushed to the scales of the infinitely small, so that it was impossible to know the true surface of the circle. Differential calculus solved the problem of "squaring the circle".

5.1.2. *A research organization independent of biblical dogmatism*

In 1739, the young academician Leclerc asked to be transferred to the botanical class of the Academy to be elected "Associate". A few months later, his friend Charles-François du Fay, member of the Academy and Superintendent of the *Jardin royal des plantes*, proposed to take over from him. The matter was heard with Secretary of State Maurepas, and Louis XV appointed Georges Leclerc as the new Superintendent of the *Jardin royal parisien*.

The Garden was created in 1635 by an edict of Louis XIII. His physician Guy de la Brosse had managed to convince him to cultivate medicinal plants in Paris, and in greater variety than in Montpellier where the tradition had been maintained since the reign of his father, Henri IV. Henri had envisioned this project for Paris but it did not succeed (Vons 2016). This garden would have only pedagogical virtues, "to demonstrate to scholars and other interested people..." and to teach the less well-off Parisians to choose the right medicinal plants. The project had been presented to Louis XIII in 1618 by his physician Jean Riolan – a young man when he held the Chair of Anatomy, Botany and Pharmacy at the *Collège royale.* For his classes, he wanted to establish a garden of his own at the University of Paris, and therefore asked that it be close to the *Collège.*

The Faculty of Medicine of the University of Paris had a monopoly on teaching surgery, medicine and apothecary and was concerned about Guy de la Brosse's project. He was a libertine intellectual, hostile to the care given abusively to kings (Kahn 2007) and influenced by the pharmacology of Paracelsus (1493–1541), imprinted with alchemy, and galenic medicine (from Galen), which did not recognize the great blood circulation demonstrated by Harvey since 1628 (the heart is not a filter between arterial and venous blood, it is a pump that sends blood to the lungs and then to the organs with a well-sealed wall between the two ventricles). De la Brosse did not hesitate to qualify the Parisian Faculty, which bleeds too easily, a "bloodthirsty sect". We can recall the insults of Jean Riolan and his friend Guy to William Harvey, ironically referring to the "great circulation" by calling Harvey a "circulator", which in Latin means "itinerant charlatan doctor".

Jean Riolan was appointed Catherine de Medici's first physician in 1632 and was to follow her in exile to London until her death in 1648. The Chair of Anatomy, Botany and Chemistry disappeared. For his part, Guy de la Brosse succeeded in the *Jardin des plantes médicinales* in being independent from the University and taught anatomy, botany and chemistry with three separate chairs. In order to avoid tensions with the powerful families of physicians who had been passing on privileges within the university for generations, Louis XIII found a compromise. The *Collège royal* founded by Francis I in 1530 was under his authority; he appointed the professors, who addressed the public in French, not in Latin. The courses were not degree courses, they did not train doctors or an intellectual elite; it was a form of popular culture. The chairs were addressed to the Parisian people, for them to better learn and understand the use of plants, in the popular language of French and not the Latin language reserved for people of letters and other academics; their vocation was popularization.

It so happens that a philanthropic herbalist and pharmacist, Nicolas Houël, owned a medicinal garden on the left bank of the Seine with a house for the apothecary (the pharmacy) and the seed bank. The estate was not far from the *Collège de France*.

Louis XIII therefore acquired it and created the *Jardin royal des plantes médicinales* with a popular educational mission and three chairs. The professors of the Medicinal Garden were thus holders of royal chairs, but they did not have the status of university professors because their courses did not confer degrees any more than those of the *Collège royal*. Parisian students were free to attend, but they were not obliged to follow them in order to obtain their doctorate in medicine, chemistry and apothecary.

The function of Superintendent was associated with the title of King's physician and the holder of the Chair of Garden Anatomy needed to be a physician of the Faculty of Paris. The superintendence was therefore entrusted to Charles Bouvard, Regent Doctor (Professor) and Dean of the Faculty of Medicine of Paris. He was also the holder of the Chair of Medicine at the *Collège royal*. Guy de la Brosse was only the Intendant.

Placed under royal supervision, the monopoly of the doctors of the Sorbonne was thus tempered because the king authorized the professors of the *Jardin des plantes* to invite professors from other universities, like the one in Montpellier, which taught about great blood circulation. Thus, the dogmatism of the Faculty of Medicine of Paris was circumvented by this clever distribution of duties and roles. Future doctors were not obliged to listen to lessons on the great circulation, but they could follow them in Paris thanks to the Chair of Anatomy of the *Jardin royal des plantes médicinales*, because Bouvard would not have opposed Guy de la Brosse, Intendant appointed by the king.

The creation of three chairs, *at* the *Collège de France*, and not *from* the *Collège de France*, thus marked a first step toward the intellectual independence of naturalists from a decision of the king. Teaching could distance itself from scholasticism, galenic medicine and the alchemical pharmacopeia of Paracelsus.

Since Louis XIII, the sovereigns kept at the Palace of Versailles all kinds of natural curiosities offered on different occasions, as well as shells, vases or semi-precious stone furniture. Louis XV decided in 1713 to transfer these curiosities to the castle of the *Jardin médicinal parisien*, which then took the name of the *Jardin royal des plantes*. These collections joined the numerous medicinal plants, drug pots and utensils necessary for pharmacopeia, as well as the *Droguier du Jardin du Roy*, composed of gems, salts, "earths" and various minerals. The establishment changed its statutes and became an institution dedicated to natural history research. The superintendence therefore no longer required the title of first doctor of the King, a graduate of the University of Paris. It was a chemist, Charles-François du Fay, who was appointed to this office to put in order this bric-a-brac of minerals, shells, vases and furniture, although this task fell to an administrator and not to a professor of chemistry. However, with the management of his domain, his woods and his scientific training, Georges Leclerc had the ideal profile, endowed as he was with a scientific culture and well-structured thinking, as his innovative article on probabilities and his taste for botany attests.

Georges Leclerc, Count of Buffon, began his Parisian career at the age of 32 years, in charge of putting this motley collection in order. As he was neither a physician, nor a chemist, nor an apothecarist, he did not hold a chair and therefore never taught at the *Collège de France*. He was in charge of classifying all these specimens, which may have contained snakes for their venom for therapeutic purposes, as well as salts. He needed to organize their study according to scientific rigor (measurements, geometry, statistics and experiments) and supervise their teaching with the three chairs. Leclerc always had a fabulous ability at research and at disseminating knowledge in French, both to the general public via the *Collège Royal* and to the European community of naturalists via the *Académie des sciences* and the publication of his dissertations in Latin, which was the international language.

The strategy of Georges Leclerc, who now signed himself "Buffon", consisted of highlighting the heritage of the royal family and enriching it, so that the Cabinet would be worthy of the king, not only in France but also in Europe; indeed, the most prestigious private collections were not lacking (Lamy 1930). Three years later, in 1742, he surrounded himself with the skills of a physician with whom he could work in peace, without rivalry, one who was hardworking and rather submissive. This doctor was Louis Jean-Marie d'Aubenton (1716–1799) who would become the master of the young Jean-Baptiste Lamarck. He was a distant cousin of Buffon from Montbard; he followed anatomy courses at the chair of the *Jardin des plantes* with a quiet, meticulous, precise and rigorous temperament; he was the ideal candidate. D'Aubenton (Daubenton) accepted the position of *Garde des collections* as well as that of demonstrator of the Chair of Anatomy held statutorily by Buffon, who did not teach. He was given the task of classifying the collections and making them accessible to the public by setting up galleries.

The pair worked and the influence of the Royal Garden aroused many material gifts in the fields of geology, mineralogy (including fossils), botany and zoology, with all kinds of animal species, shells, insects, skeletons of reptiles, birds and mammals, apes and humans included. Daubenton undertook the study of species according to a systematic method that laid the foundations of comparative anatomy. The first publications appeared as early as 1749 with quality reproductions. Buffon signed his works; Daubenton appeared more as a "research assistant" who accomplished the most laborious tasks and who could add his name to them.

5.1.3. *The history of the Earth as a premise of the natural history of Man*

Aristotle's natural history of animals remained a major reference, enriched since Antiquity by the publications of naturalists, whose faithful observations, fertile imaginations and notorious falsifications had to be taken into account, as Georges Cuvier did not fail to point out in his lectures at the *Collège de France.*

Georges Buffon published a "Natural History", a "Theory of the Earth" and a "History of Man" combined into one book, general and particular, with the description of the King's Cabinet. It is an encyclopedia which appeared in 36 volumes between 1749 and 1789. The last eight volumes were published posthumously by Lacépède (1756–1825) until 1804. The encyclopedia concerns minerals and is limited to quadrupeds and birds. The first three volumes appeared in 1749: *De la manière d'étudier l'histoire naturelle, la Théorie de la Terre and Histoire générale des animaux et Histoire naturelle de l'homme*. The *Histoire générale des animaux* includes Daubenton's description of about 200 species of quadrupeds. They were a huge success with the public.

The *Nomenclature des singes* was published in 1766 in volume XIV of the *Histoire Naturelle des Quadrupèdes*. We will come back to it later.

> The first volumes were not examined by the censors nor by the *Académie des sciences* of which Buffon was a member. As early as 1751, the Sorbonne was concerned about certain proposals. But the position of the intendant of the King's Garden, having his work published by the royal printing house and possessing a personal fortune, is more solid than that of Diderot. (Delon, *À propos de l'œuvre*, n.d. read online, author's translation)

An edition from 1838 published the complete texts as well as additions and corrections, with the letters of *Messieurs les Députés et Syndic de la Faculté de Théologie de Paris* dated 1751:

> Sir, we were informed by one of us, on your behalf, that when you learned that Natural History, of which you are the author, was one of the works that were chosen by order of the Faculty of Theology to be examined and censored, as containing principles and maxims which are not in conformity with those of religion, you declared to him that you had not intended to deviate from them, and that you were prepared to satisfy the Faculty on each of the articles which it would find reprehensible in your said work; we cannot, Sir, give too much praise to such a Christian resolution; and in order to put you in a state of execution, we are sending you the proposals extracted from your book, which seemed to us contrary to the belief of the Church. We have the honor to be in perfect consideration, Sir, Your most humble and most obedient servants. (Buffon, *Histoire naturelle*, 1838, author's translation)

The censorship exposed 14 proposals "which appeared reprehensible". Theologians attacked its statement that mathematics and physical verification were the primary means of establishing knowledge of nature. Moreover, they systematically rejected the major theses that make the *History of the Earth* a key work without equivalent in the history of geology: the Earth had a history in six stages with a period during which unknown species, known as "lost species", moved between Europe, Asia and America, which formed a single continent.

Fossils, both marine and vertebrate, were well known to European naturalists, the Garden's collections were well-endowed, and Buffon surveyed the rolling countryside of the Dijon region enough to form his own opinion, observing the Langres mountain range, the highest to his knowledge, as this range separates the two great river basins, the one that flows onto the western side of France, and the other that reaches the Mediterranean Sea. Obviously Buffon did not visit the Alps, but he reported a remarkable phenomenon of ice covering valleys: glaciers. He did not doubt that seas occupied areas covered with hundreds of meters of sand, mud and rocks, more or less indurated, that some deposits formed the superimposed horizontal layers that are found between hills, that some are crossed by planes (faults) that are found for miles. Their age and their distance from the oceans called into question the age of the Earth. The mineralogical collections included volcanic stones such as obsidian; Buffon had become aware of this during a trip to Italy (Condorcet 1790). Active European volcanoes were known in Italy in 1538 at Campi Flegrei, in 1564 in the Azores with the Agua de Pau and in 1630 in Furnas (Portuguese islands in the Atlantic Ocean). This viscous molten matter that froze into silica led him to conclude that the Earth kept in its depths the traces of a celestial past that was much hotter, even incandescent.

In his *Theory of the Earth*, Buffon assumed that the Earth's mass was composed of vitrified silica that had undergone a molten phase. An admirer of Newton's work, he nevertheless rejected the theological dogma of the first movement supposed to have animated the solar system. Theories on its origin evolved with the theory of the protosolar nebula of Swedenborg (1688–1772), Immanuel Kant (1724–1804) and Pierre Simon Laplace (1749–1827). It was up to him, the Superintendent of the Royal Collections and member of the *Académie des sciences*, to make this known. For Buffon, discussing mechanical shock:

> a comet falling into the body of the sun, will displace and separate some parts to which it will have imparted an impulse movement … so that the planets would have once belonged to the body of the sun, and been detached from it. (Buffon, *Theory of the Earth*, 1749)

The hazardous event became a factor in the physical organization of nature, namely, the meeting of two independent trajectories. Chance became necessary to understand the course of the history of the cosmos, the solar system and life on the Earth. Broken by chance and the almost infinite variety of landscapes of the Earth's crust, formed of folds, reliefs, caves and underground galleries, with its varieties of rocks and sands and frozen lava, the variety of the sea beds described by explorers, the impermanence of the soil exposed to the seasons and earthquakes, there was no doubt that the Biblical Flood was a legend passed down from the very first human populations. These primitive people were frightened by a composition of the Earth still relatively primitive, because still hot, with erupting volcanoes, earthquakes, catastrophic floods (tsunamis), and sought refuge in the highest reliefs of Central Asia where the first couple would have been born.

In 1768, Buffon had a large forge built on his own land and then conducted an experiment to calculate the age of the Earth by white-heating all kinds of balls composed of different minerals, or rocks, of varying diameters. Then he measured the cooling time of each ball. The results extrapolated to the diameter of the Earth gave figures up to 10 million years. The tables were published in a new book, entitled *Les époques de la nature,* in which *he does not mention this last result. He explained this later, holding such an ancient age to be unreasonable; it is unknown whether he really adhered to this self-censorship or whether he did so thanks to the Sorbonne, being rather less daring than Galileo.* The censorship of the theologians was right, however, for he was determined to give an age of 75,000 years for the Earth, specifying that:

> One will feel that this enormous duration of seventy-five thousand years is not yet enough for all the great works of nature. (Buffon, *Histoire naturelle, générale et particulier*, Part 6, 1785)

He then proposed a succession of very long periods from the age of molten lava to that when the continents united and then split up. It was a true tectonics of the Earth that he put in place, an already very audacious posture.

Buffon explained that the Earth not only had a past, but the planet also had an irreversible becoming that was not random: this future was the one imposed by natural and cosmic determinism due to the cooling of its still molten core (volcanoes attest to this). To the cycle of its rotation around the sun was added a new dimension of terrestrial time, that of a time irreversibly oriented toward a stable state. It was not a duration, like the development of a living organism, but it was already the prospect of an "evolution" oriented on a planetary scale and on the scale of millions of years (he knows this potentially), determined by physical laws: the degradation from gaseous and hot (nebulous) to viscous and hot, then solid and cold.

The mineralogy collections of the *Jardin des plantes* preserved fossils of invertebrates (ammonites, oysters) and much rarer, very large mineralized bones. The innumerable "shells" in quarries or in the outcrops of roads and rivers were stratigraphic geology. They were markers that made it possible to situate the layers in relation to each other, parameters of the physics of the globe that are not the subject of a natural history. Their history would require a vast program of anatomy compared with current species (Balan 1979). This would therefore be Georges Cuvier's program for vertebrates, but not before the 19th century.

The theory of the cooling of the Earth allowed Buffon to build an account of the origins of life, the first vertebrates and the first humans. The *Jardin des plantes* received fossil bones of rhinoceroses and hippopotami from Europe, as well as very large elephant tusks from Siberia, Tartary and North America, completed with a huge molar that was sufficiently distinctive to reveal a "lost" species, in his own words. On the basis of the facts, he admitted the existence and disappearance of unknown genera. These lost species would be called "mastodons".

The history of the Earth was summarized in seven major periods that spanned at least 75,000 years. The first lasted between 5,000 and 10,000 years: it corresponded to a mass of molten silica that swelled at the equator and flattened at the poles under the effect of the Earth's rotation on its axis; the second corresponded to the first phase of the very long cooling, with the condensation of water vapor and the formation of an ocean that covered the Earth; it lasted 25,000 years. The third period corresponded to the appearance of life in the oceans, attested by the deposits of invertebrate shells and terrestrial plants (petrified in coal); it lasted between 15,000 and 20,000 years:

> All we can say today is that some of the monuments (fossils) show us the existence of certain terrestrial and marine animals whose living

> analogues we do not know (...). Nature was then in its first strength, and worked organic and living matter with a more active power in a warmer temperature (...). By fertilizing the seas, Nature was also spreading the principles of life on all the lands that water had not been able to overcome or that it had promptly abandoned. (ibid.)

The fourth geological period would have lasted 10,000 years. It corresponded to the retreat of the sea, with the formation of reliefs on the continents due to the hollowing out of valleys, because the mass of oceanic water was lowered following the collapse of its ocean floor caused by the cooling of the core, and thus the retraction of the superficial crust. Volcanoes became active and contributed to the thickening of the continents:

> they are natural effects produced by a natural cause, i.e. by the action of cooling on molten matter when it consolidates on the surface. (Buffon, *Histoire naturelle, générale et particulier*, 1785)

The fifth period was only 15,000 years before the present. It began with a cooling threshold sufficient to allow the appearance of terrestrial vertebrates in the northern hemisphere, and not in the southern hemisphere where the ocean dominated most of the Earth's surface and where the land masses in the tropics and the equator were still unfit for life. Indeed, the fossil species of these genera corresponded to the current species adapted to the tropical climate. Buffon deduced that the planet was then much warmer toward the poles and uninhabitable at the equator, still unfit for life because of the Earth's slower cooling. The terrestrial fauna thus appeared spontaneously in the great diversity of environments of an immense continent that occupied the northern hemisphere, from America to Siberia via Europe. Then the northern continent collapsed between Europe and America. Buffon thus introduced a new science in geology, the tectonics of the globe. Tectonic movements are only perceived vertically: the idea of tectonics in the direction of the Earth's curvature would only be definitively accepted at the end of the 1960s with plate tectonics, allowed by the formation of convection currents of lava, a consequence of the cooling of the surface (thermal gradient from the depths).

When the global temperature had sufficiently decreased, the sixth period arrived with an intense cooling phase that drove elephants and other Eurasian species to the tropics, which became habitable. The American elephants disappeared, while South America was spontaneously populated with new species. The poles were covered with ice; deer and polar bears flourished. Finally, the seventh period saw the spontaneous appearance of human beings, "in the center of the continent of Asia, from the fortieth degree of latitude to the fifty-fifth (...). Siberia and the Tartary" (Buffon 1785).

5.1.4. *Man is the last "internal mold" created on the Earth*

Buffon does not go into detail concerning the times since the appearance of terrestrial species; we have to imagine these by ourselves. Man appears last on a still warm Earth, with a profusion of volcanoes and associated earthquakes.

Man forms a single species that quickly conquered the continents thanks to a technical "superiority of intelligence" offered by his brain and skillful hands. Geographical diversity, reported by navigators and explorers, illustrates minimal varieties of adaptation to the climate. With its civilizations and domestic conquests, the human species even became a factor on the scale of the Earth's cosmic forces, since the heat released by its creativity reversed the Earth's tendency to cool; Buffon already had an idea of the Anthropocene Era!

His words are strikingly modern:

> Let us therefore assume the world in peace and see more closely how the power of man could influence the power of nature. Nothing seems more difficult, not to say impossible, than to oppose the successive cooling of the earth, and to warm the temperature of a climate; yet man can and has done it. Being at the same latitude, Paris would be as cold as Quebec City if France, and all the countries around it, were as devoid of men, as covered with woods, as bathed by the waters as Canada's neighboring lands are. To sanitize, clear and populate a country is to return heat to it for several thousand years: and this prevents the only reasonable objection one can make against the real fact of the cooling of the Earth. (Buffon 1785)

These unformalized geophysical properties of Buffon's time were formalized at the beginning of the 19th century with the three principles of a new science, thermodynamics. It developed in the 20th century and enabled progress in the life sciences without succeeding, until the first quarter of the 21st century, in clarifying the mechanisms of the first cycle of cell reproduction.

His sense of observation and logical reasoning allowed him to propose a richly documented and fruitful history of the Earth for the intelligibility of the sensitive and observable world. However, as soon as the question of life was raised, i.e. organized systems that reproduce themselves, his naturalistic thinking remained vague and that of the mathematician lost its power. His system of the origins of life was neither a scientific theory nor a form of metaphysics. It was an unclassifiable chimera, wanting too much to reconcile the history of the Earth and animals with the Genesis of the Bible. That said, no mathematical model, however complex it may be, has yet succeeded in producing life in the laboratory, i.e. a cell cycle that registers by itself as it is formed.

Inevitably, the natural history of animals and the human species gave rise to a position on the doctrines of species generation and therefore on the notion of species. However, before arguing his theory of generation, Buffon was forced to pledge allegiance to the theological authority of the Sorbonne by acknowledging that it was by the "book of revelation" (a knowledge transmitted by supernatural means accepted as such), and not by human reason, that intelligence accesses the explanation of the world. This submission to authority distinguishes him from Descartes and Leibniz. Georges Buffon had the makings of an entrepreneur, an engineer and a historian, but not that of a theorist and even less that of a metaphysician. In short, he did not need to understand the origin of the beginning; his reasoning gave way to his belief in the supernatural inspiration of the Genesis story:

> We must therefore say that, being formed of earth and composed of dust, we do indeed have a common relationship with earth and dust that binds us to matter in general, such as extent, impenetrability, gravity, etc. (…) but as we do not see these purely material relationships, we cannot say that they are part of our being, it is therefore the organization, the life, the soul, which properly makes up our existence (…). We therefore exist without knowing how, and we think without knowing why (…). The idea of reducing the explanation of all phenomena to mechanical principles is certainly great and beautiful, it is the boldest step one could take in philosophy, and it was Descartes who did it; but this idea is only a project; and is this project well-founded? Even if it is, do we have the means to carry it out? (…) To want to admit in the matter only the qualities that we know about, seems to me a vain and ill-founded pretension. (Buffon, *Histoire naturelle*, third volume, 1769, author's translation)

The objective being to reconcile the history of the Earth and of animals with biblical dogma, Buffon proposed a system without worrying about the unresolvable problems that were added to each scaffolding of his reasoning. The result is a curious intellectual chimera, the exact opposite of what rational and scientific minds have been trying to put in place since the Renaissance. The keystone is his idea of the "internal mold" which adds more shadows than it illuminates to the physical understanding of generation:

> It thus seems certain to us that the body of the animal or vegetable is an internal mold which has a constant shape, but whose mass and

> volume can increase proportionally, and that the increase, or if one wants, the development of the animal or vegetable, is only done by the extension of this mold in all its external and internal dimensions; that this extension is done by the intussusception[2] of an accessory and foreign matter which penetrates into the interior, which becomes similar to the shape, and identical with the matter of the mold (…). The internal mold acts by the play of penetrating forces of which we will never have a clear idea. (Buffon, *Histoire naturelle*, 1749, author's translation)

According to Fischer:

> The internal mold for Buffon corresponds to a property of Nature in the same way as the force of attraction. (Fischer, *Buffon et les théories de la génération au 18ème siècle*, 2003, author's translation)

In reality, it is a chimera; his "model" is unsustainable in metaphysics as well as in physics. Buffon approaches the generation of species by taking up the different doctrines of animalcules in semen, which are in reality, according to him, "living organic molecules":

> Maybe I could recognize the living organic parts, from which I thought all animals and plants originated. My first suspicion was that the spermatic animals (the animalcules), which we saw in the semen of all males, could well be only a very large number of living organic molecules: but if so, the semen of the female must contain roughly similar living organic molecules. I therefore thought of examining the seminal liquors of males and females under the microscope and I made a plan of experiments; the semen reservoir of the females could well have been the glandular body cavity, in which Vallisnerie and the others (including de Graaf) had uselessly searched for the egg. (Buffon, *Expériences au sujet de la génération*, 1830, author's translation)

The "internal mold" was inspired by the footprints of fossil ammonites in hardened clays: the extinct animal left an external shape that was in contact with the middle (the edge which was a surface), and an internal volume shape in the hollow; "a massive surface", Buffon wrote. The forces belong to the physical world, finite and mortal, yet the creatures of the Bible are eternal. Buffon therefore added an additional condition: the "living organic molecules" are immortal, they are part of

2 Mode of growth of living cells by assimilation of new elements in the mass and which opposes them to crystals, which grow by superposition.

the design of the mold; they return to the environment at the death of the organism and thus remained available:

> The concept of organic molecules corresponds to an intellectual logic that fits it: they are immortal since the organized being, which is an assembly of these organic molecules, returns them to nature after its death. (Fischer 2003, author's translation)

The term "living" means "that which reproduces", but, in Buffon's mind, it has a different conception. The mold of the species has an external surface open to the environment and a shaped internal surface that organizes "organic molecules" drawn from the environment. This organizing surface is potentially life, it is potentially there. It is immutable, unalterable, and therefore did not have the properties of the physical world, which had a beginning and an end like the history of the Earth. It was not the molds that changed, it was not the life specific to each mold that disappeared; life as potentiality never disappeared. The change came from the immediate material conditions, it was the quantity and quality of the "living organic molecules" in the water or in the wet mud that allowed molds to actualize themselves. If the physical conditions changed, then the old species disappeared and molds that remained "inactive" gave rise to new species by taking over the "living organic molecules". Life emerged when the organizing surface was filled with these "organic molecules". Balan called them "metaphysical molecules".

Buffon had the opportunity to verify that sperm and oocytes were "animalcules". This study was conducted with Daubenton during the visit of a specialist in microorganisms, John Needham (1713–1781), equipped with a high-performance microscope. They conducted *in vivo* fertilization experiments, followed by dissections. Buffon confirmed that the oocytes of a female dog were composed of these "living organic molecules" (Buffon 1769).

His theory is not an explanation of the origins of life and it posed a problem for epistemologists because it was contradictory. Indeed, once his postulate was established, he did not propose a dynamic physics to conceive how this "internal mold" proceeded with these "living organic molecules". The term "living" did not refer to the forces of the molecules, precisely because they were not dynamic, they were not attractions between different atoms that self-organized and reproduced themselves. They were sucked in by the mold; in short, it was the organizing inner face which was the conductor and which associated and oriented the molecules to form the organs.

Man's animalcules and woman's oocytes were therefore finite assemblages of "living organic molecules", and it was at the moment of fertilization of the oocyte by the animalcule that the internal mold became active thanks to the pooling of the "living organic molecules". The embryo began to form under the organizing effect of the internal mold as it sucked the other "organic molecules" from the maternal matrix, then through the umbilical cord:

> The formation of the fetus is thus done by the union of the organic molecules contained in the mixture which has just been made of the seminal liquors of the two individuals: this meeting produces the local state of the parts, because it is done according to the laws of affinity which are between these various parts, and which determine the molecules to be placed as they were in the individuals who provided them; so that the molecules which come from the head, and which must form it, cannot, by virtue of these laws, be placed elsewhere than near those which must form the neck, and that they will not go to be placed near those which must form the legs. (Buffon 1838, author's translation)

The "internal mold" was thus a chimera for a metaphysician; an immortal, invisible form could not be endowed with attractive physical properties and Buffon's reasoning lacked the metaphysics from which he subtracted himself in order to avoid pushing the reflection to the origin of movement (Descartes) and thus to the origin of a force (Leibniz), which attracts or repels the atoms associated by affinity (Newton).

The dilemma is considerable. As a naturalist, Buffon took a reasonable position, because the doctrine of the Leibniz monads implied, for the generation of the human species, for example, that all future human beings were in all the spermatozoa and oocytes of the men and women who populated the Earth in the 18th century, including in his own spermatozoa. This was obviously inconceivable to Buffon.

In his system, the "living organic molecules" are part of the creation of the "internal mold"; they are created and seeded into the waters and puddles, destined to be sucked up by "internal molds".

Space and time are still dissociated in Georges Buffon's thinking. Time is geological and irreversibly determined by the cooling of the Earth's core and its side effects; time is a consequence, and it is not an organizing duration in the way that microbe-sized organisms were. The cosmos is not a duration in itself, which organized itself irreversibly with its forces of attraction and repulsion.

For Buffon, the sperms of man and woman are therefore not monads that contained the form (topology in mathematics). No preformation was present in the animalcules and oocytes, the "internal mold" and its "organic molecules" belonged to another level of reality than that of physics: it was the invisible reality of metaphysics.

5.1.5. *The species according to Buffon*

With Buffon's "generation theory" posed, we can understand his definition of the species. The naturalist aligned himself with Aristotle's logic by starting from the best known of four-legged animals, Man, objectively constrained to observe a degradation within the limits of the same principle of composition, which he called "design": a plan of organization of forms in the space of the system, that of an oyster for example. It was a question of form and space in the organization of the animal or plant, not of a plan purposefully organized by the Creator following the geological time scale. From Man to oyster and from oyster to small animal, the organization follows a degradation.

Buffon then reasoned about the only possible and known way that generated an organization and which was thus reproduction by the fertilization of seeds:

> Do the donkey and the horse therefore originally come from the same stock (…) where one could attribute the very slight differences to the very ancient influence of climate, food, and the very fortuitous generation of small, half degenerate wild horses? Are they, as those who name them say, of the same family? Or are they and have they not always been different animals? This question, whose generality, difficulty, and consequences physicians will well know, and which we thought we should deal with in this article because it arises for the first time, relates to the production of beings more closely than another, and requires, for clarification, that we consider Nature from a new point of view. (Buffon, *Histoire naturelle. Les Quadrupèdes*, 1838, author's translation)

He notes that quadrupeds are built with almost the same bones, with teeth, sense organs, a circulatory system and lungs and a reproductive and digestive system; the differences are in their spatial arrangement and number. Horses gallop with a single phalanx while apes walk with their palms or five fingers, cetaceans move with their powerful fins made up of the bones of a wrist and five fingers:

> If, in the immense variety presented to us by all the animated beings that populate the universe, we choose an animal, or even the body of man, to serve as a basis for our knowledge, and to bring to it, by way of comparison, the other organized beings, we will find that, although all these beings exist alone, and all vary by differences graduated to

> infinity, there exists at the same time a single primitive and general design that can be followed very far, and whose degradations are much slower than those of the figures and other apparent relationships (…) because there is in the very parts that participate most in the variety of the external form a prodigious resemblance that necessarily reminds us of the idea of a first design, on which everything seems to have been conceived. (ibid., author's translation)

The term "degradation" was intended to be understood as such. To better clarify his point, Buffon imagines an adult human skeleton transforming into a horse skeleton:

> Take the skeleton of man, bend the bones of the pelvis, shorten the bones of the thighs, legs and arms, lengthen the bones of the feet and hands, weld the knuckles together, lengthen the jaws by shortening the frontal bone, and lengthen the spine of the back; this skeleton will cease to be that of a man's remains, and will be that of a horse. (ibid., author's translation)

He could have written: saw a human skeleton in half through its middle. Move up the part of the base of the skull that is vertical and you will then see the lower jaw joint move upward, and its facial part follow the opposite movement, that is to say, a lengthening toward the front developing prognathism. Simultaneously, the spine becomes horizontal, the upper limbs become anterior and the lower limbs become posterior. In the end, you will not have a horse, but a four-legged prosimian, which, in steps, resembles the great ape, then the little ape.

For classification, or nomenclature, Buffon's view is simple: the only objective natural unit is a population of interbreeding individuals who share a single unit of anatomical organization. Each individual is different in that it is unique, but all are related in that they are interbreeding; and this unit forms a species regardless of the number of generations that have passed:

> The species is thus an abstract and general word, which exists only by considering Nature in the succession of times, and in the constant destruction and equally constant renewal of beings. It is by comparing the Nature of today with that of other times, and the present individuals with those of the past, that we have taken a clear idea of what is called species, and the comparison of the number or resemblance of individuals is only an accessory idea, and often independent of the first one. (ibid., author's translation)

Beyond the species, categories are only creations of the human mind, which needs to prioritize and build mental "drawers" to put its samples in some kind of order. It is a pyramidal vision that has functioned for centuries on the principle of the first order, scattered in sub-orders, with subordinates, themselves framing classes, which themselves frame superfamilies and so on down to the individual. Order maintains the system. The animate world must be ordered, and it is therefore ordered according to a static hierarchy: order first, then classes, and in the classes families, and in its families genera, and these genera are composed of species, which themselves break down into subspecies with their varieties, and at the end of the day, individuals:

> Although the march of nature is done by nuances and often imperceptible degrees, the intervals of these degrees or these nuances are not all equal, to a great extent (...) we must not forget that these families are our work, that we have made them only for the relief of our spirit; that if it cannot understand the real continuation of all beings, it is our fault, and not that of nature, which does not know these alleged families, and indeed contains only individuals. (ibid., author's translation)

The definition of the species is:

> Neither the number nor the collection of similar individuals, it is the constant succession and the uninterrupted renewal of these individuals that constitute it. (...). The species is thus nothing other than a constant succession of similar and reproducing individuals. (ibid., author's translation)

Regarding humans, Buffon concludes:

> There is only one and the same species in man (...) If the black man and the white man could not produce together, if even their production remained infertile, then there would be two quite distinct species. But this very supposition is disproved by the facts; and, since all men can communicate and produce together, all men come from the same stock and are of the same family. (ibid., author's translation)

What then is the origin of a species?

> In our hypothesis the watery, oily and ductile (soft) parts, which were to enter into the composition of the organized beings, fell with the waters on the northern parts of the globe, much earlier and in much greater quantity than on the southern parts. It was in these aqueous

> and ductile materials that living organic molecules began to exert their power to shape and develop organized bodies. (ibid., author's translation)

> Every production, every generation, and even every increase, every development, presupposes the competition and union of a great quantity of living organic molecules; these molecules, which animate all organized bodies, are successively employed in the nutrition and generation of all beings. If all of a sudden the greater part of these beings were to be suppressed, new species would appear, because these organic molecules, which are indestructible and always active, would come together to compose other organized bodies; but being entirely absorbed by the internal molds of the existing beings, no new ones can be formed (in the northern hemisphere) … whereas, on the contrary, in the lands of South America, where the large animals of the north could not penetrate, the living organic molecules, not being absorbed by any animal mold already existing, will have joined together to form species that do not resemble the others (…). These two formations (North and South America) are made in the same way and by the same means; and if the (northern species) are first, it is because the fecundity of the Earth, i.e. the quantity of living organic matter, was less abundant in these southern climates. (ibid., author's translation)

Jacques Roger summarized Buffon's geological and geochronological analysis:

> He gave geology the status of a historical science that it will retain throughout the 19th century. From then on, it will be imperative to identify more precisely these fossils which are used for the relative dating of rocks and which inevitably lead to the notion of evolution. Directly or indirectly, through Cuvier or Lamarck, Buffon's influence will be considerable. But what is perhaps most modern in him is this "philosophical spirit" that allowed him to get rid of archaic but tenacious prejudices, and which will survive him, to give geology the status of a "natural" science, and to pursue to the end the study of the fundamental questions that this science could ask. (Roger, "La place de Buffon dans l'histoire des Sciences de la Terre", 1988, author's translation)

Buffon thus happily admitted that unknown species inhabited a period of the Earth with climatic, tectonic and geochemical conditions different from those of living species. He amply recognized the possibility of anatomical variations induced by climatic and environmental conditions over generations and he also recognized that these varieties were due to variations in the Earth conditions; they were assimilable by the "internal mold" within the limits of its extension. He recognized that the disappearance of species and their replacement by new species were caused by the geophysical determinism of the Earth, the center of which tended toward the solid state by shrinking. Everything was in place so that geochemistry, geotectonics and geothermics on a planetary scale could be understood as the causes of the extinction of unsuitable species and the factor of the appearance of new species. Diversity or microevolution was recognized. The natural history of the Earth during the great periods allowed microevolution within the limits of the molds. On the other hand, under no circumstances could the transitions between the last great terrestrial periods constrained by cooling have modified the molds. The division is all the more striking since Buffon did not say how the vertical anatomy of Man had a relationship with the last stage of the Earth's cooling. For several generations, naturalists had suspected that apes, with their so-called "anthropoid" features, had a terrestrial history in common with humans. Buffon was necessarily aware of this; he knew that the Anthropoid had four palms and could occasionally be bipedal, whereas humans had only two palms and two feet and were exclusively bipedal, yet the morphologies were radically distinct.

In his intellectual essay on the process of "degeneration" as the only process capable of creating a new genus, Buffon concludes that this theory is implausible. He therefore rejects the application of the ape principle to humans, since degeneration was a loss and not a gain; it goes without saying that, in this case, the ancestor would be the human, with its big brain, and not the ape.

> One could also say that the ape is of the family of man; that he is a degenerate man; that man and the ape had a common origin like the horse and the donkey (…) But no, it is certain by the revelation that all animals also participated in the grace of creation; that the first two of each species and of all species came out all formed from the hands of the Creator; and one must believe that they were such then approximately that they are represented to us today by their descendants: moreover, since we have been observing nature from the time of Aristotle to ours, we have not seen any new species appear. (Buffon, *Histoire naturelle. Les Quadrupèdes*, 1838, author's translation)

Georges Buffon never considered the ape as a degenerate type of the human; he recognized an identity and integrity as worthy as any other "mold created by the Supreme Being".

He also laid down great principles that did not lose their relevance until the 21st century: the subordination of external parties to internal parties.

5.1.6. *A fundamental principle: the subordination of external parties to internal parties*

The most constant parts are the most essential – such as the brain. Buffon insisted on the pre-eminence of the sensory organs and distinguished thought from instinct. He insisted on the forces that animate movement at all scales of organization: not only the stars (Newton, gravitational force), not only the mobility of organisms (muscles, innervations), but also at the scales of their organization imperceptible to the naked eye.

> The real motivations of our organization are not these muscles, veins, arteries, nerves; there are *inner forces* in the organized bodies, which do not follow at all the laws of gross mechanics that we have imagined, and to which we would like to reduce everything: they have reappeared with more brilliance than ever, in gravitation, without chemical affinities (Newton), in the phenomena of electricity, etc. ... But, in spite of their obviousness and universality, since they act inside, since we can only reach them through reasoning, since they escape our eyes, we find it difficult to admit them, we *always want to judge them from the outside, we imagine that this outside is everything*: it seems that we are not allowed to penetrate beyond, and we neglect everything that could lead us there. (Buffon, *Histoire naturelle de l'Homme*, 1839)

Calculations of the ages of the Earth reached astronomical values with the geologist and Catholic priest, Jean-Louis Giraud-Soulavie (1751–1813); his results suggested several 100 million years. His work was published in seven volumes (Giraud-Soulavie 1780–1784) and followed Buffon's reasoning, which divided the Earth's history into five periods. He conceived an evolution of plants and fauna under the effect of changes in the environment. In turn, he was censored by the Catholic Church, which forbade the publication of the last volumes.

This final resistance will be countered, due to the French Revolution, by Jean-Baptiste Lamarck, who had the audacity to accord the quadrumane ape and *Homo sapiens* a common ancestor, the ape having anteriority over the human. A few naturalists followed Lamarck at the beginning of the 19th century and took the risk of reconsidering the identity of humans in the natural history of the Earth and animals, confronting new and formidable difficulties. However, before embarking on this new age of science, it is necessary to make an inventory of the knowledge in primatology acquired by Daubenton and on which Lamarck based his reasoning.

PART 2

The Place of Humans among Current and Fossilized Primates

6

From Natural Curiosity Cabinets to the First Primate Collections

6.1. Introduction

The study of natural history is a necessity, which cannot be conceived without long voyages across seas and continents in order to collect specimens and compare them more easily in Europe, preserved by private individuals or rulers.

6.1.1. *Conrad Gessner (1516–1565), the first great collector of natural curiosities*

Gessner was born in Zurich. According to Cuvier, he was the first zoologist of modern times. He devoted himself to medicine and obtained a doctor's degree from the University of Montpellier in 1541. His thesis consisted of determining whether the center of emotions and sensations was in the heart, as Aristotle argued, or in the brain, as in the tradition of Hippocrates and taught by André Vésale. In an attempt to answer this question, the doctor engaged in the Latin translation of the main surgical treatises of Antiquity and concluded that the center of the emotions was in the brain (Vesalius 1555).

In Montpellier, he befriended Guillaume and Pierre Belon who had strongly influenced him since he gave himself as a program the description of all known animals. His collection included fossils, which at the time corresponded to any natural object collected in place in the earth, both minerals and organic remains (shells, bones). He was called the "Swiss Pliny". His collection was large enough for him to publish a critical catalog *Historiae animalium* (1551–1558) completed with the knowledge he had acquired. This treatise on the history of animals was considered to be the first work of modern zoology and remained the reference work for two centuries. In 1565, the year of his death, Gessner published the first work on fossils and thus the first illustrations of them, *De rerum fossilium*, which, still according to Cuvier, marked the birth of paleontology. His major contribution to the history of science was a bibliography of all Greek, Latin and Hebrew authors, the *Bibliotheca universalis* (Gesner 1545–1555).

His treatise was illustrated with woodcut plates. Aristotle's classification of anthropoids, or *simia* and *pithekos*, was divided into two groups distinguishing between apes with tails and without tails. Gessner listed seven genera of monkeys in a work dated from the 10th century, the *Nicephori Callisti Historia*: *Simiarum generibus diversis: Aegopitheco, Arcopitheco, Leontopitheco, Cynocephalo, Pane, Satyro, Sphinx*.

These genera were Aegopithecus, Arcopithecus, Leonpithecus, Cynocephalus, Pan, Satyr and Sphinx.

The drawings were not always faithful to reality: the face of the Barbary ape (Gibraltar and North Africa) which has only a relictual tail, was human. The animal was represented seated, while a new genus that he called *Papionem potuis* or Babuin was in quadruped locomotion mode with a tail stump (Gessner 1545–1555). A third illustration was drawn from an impression by Bernhard von Breydenbach (1440–1497), canon in Mainz, which appeared in 1468 in his book *Peregrinatio in Terram Sanctam*. The canon had visited Palestine, Sinai and Egypt accompanied by a wood engraver, Erhard Reuwich, from Utrecht (The Netherlands). He drew a hairy ape with a long tail, its face was a muzzle with a receding mandible (no chin), surrounded by a mane similar to that of a lion, in a vertical posture with two large udders.

This posture "humanized" the tailed ape by showing a hand holding a camel on a leash, while, with the other, it was leaning on a cane; the hands were human but, on the other hand, the feet were simian with the spread grasping thumb. Gessner repeated this illustration, maintaining the bipedal posture and the cane in the right hand, which impressed readers for a long time (Fig. 17 in Figure 6.1).

The profusion of new genera since Aristotle prompted Gessner to order the species according to a binominal classification with names in Latin. The first was the genus and the second was the species, a variation of the genus. This classification was taken up by Linnaeus. Zoologists did not know of any other anthropoids other than the Barbary and African coastal apes. The extension of colonies in Southeast Asia and Africa allowed Europeans to discover large anthropoids more similar to humans in the expression of their face and their gaze than that of a small ape with or without a tail. They were quadrupedal and arboreal, occasionally bipedal. Their behaviors and their expressions awakened curiosity, while natural history always remained that of Aristotle, enriched with very distinct families and genera, without any relationships.

Figure 6.1. *The first three illustrations of apes in Tyson (1699); Fig. 16: Bontius's orangutan; Fig. 17: Gessner's small-tailed ape; Fig. 18: Tulpius's* Homo sylvestris *© UMR 7194 – Bibliothèque de l'Institut de Paléontologie Humaine*

6.1.2. *Ulisse Aldrovandi (1522–1605), the first natural history museums in Europe*

Aldrovandi was born in Bologna in 1522. A Doctor of Medicine, he taught anatomy at the University of Bologna from 1554 and performed numerous dissections. He published a treatise on anatomy in which he described children suffering from hypertrichosis (strong pilosity) accompanied by illustrations. In 1568, he obtained the right for the university to have its own garden of medicinal plants open to the public. In addition, he had built up a cabinet of natural curiosities with 18,000 pieces which he bequeathed to the university. He was thus at the origin of one of the first natural history museums in Europe. He also donated 3,600 books to the senate of the city with nearly 300 manuscripts, to which was added his own herbarium with a wealth of more than 7,000 specimens. Aldrovandi's ambition was to publish an encyclopedia of natural history, but this was interrupted by the bankruptcy of his publisher. Cuvier remained rather critical of the value of this collection, "an immense compilation without taste or genius" but which did not put the considerable amount of information collected in its historical context: the systematics was not better thought out than in the royal collections entrusted to

Georges Buffon in 1739. It was probably in homage to his herbarium that Linnaeus gave his name to designate the species of Gessner's tail monkey, *Lucifer aldrovandi*.

6.1.3. *Jacobus Bontius (Jacob de Bondt, 1592–1631): the first wild great ape or "Man of the Woods"*

Born in 1592 in Leiden, Bontius died in 1631 in Batavia (Dutch East Indies). Born into a family of doctors, he was a doctor himself. In 1626, the Dutch East India Company entrusted him with surgical inspection in the archipelagos of Southeast Asia. In addition to an investigation into tropical diseases and exotic plant treatments, he noted the visual testimonies of natives describing naked and hairy men and women with prominent abdomens, called orangutans, or the "Man of the Woods" (Bontius 1658). He engraved and published an illustration that shows a body and face of a perfectly human woman, with locks covering the body, a swollen belly, the face surrounded by a mane as in Gessner's book, which was the reference, but without the cane. In short, this was a chimera that combined the female ape-man of Ethiopia (actually a four-legged cercopithecid) and a human body, in reference to the name given by the Indonesians, "Man of the Woods". This figure therefore had no naturalistic value (Fig. 16 in Figure 6.1).

6.1.4. *Tulpius (1593–1674), the first description of a chimpanzee*

Nicolaes Pieterz, known as Tulpius, was born in Amsterdam, the port and capital of a new colonial empire like the Alexandria of Antiquity. He was a doctor-surgeon of great reputation. He was the burgomaster of Amsterdam and must undoubtedly have had easy access to the Dutch East India Company to obtain corpses of different species brought back from the colonies by ships. Tulpius is the nickname he gave himself in reference to the tulip, a flower coveted by the elites of the prosperous city. Rembrandt made him famous with his "The Anatomy Lesson of Dr. Nicolaes Tulp".

Intellectuals were questioning the distinction between human and animal physiology at a time when René Descartes was reasoning about movement as the primary cause, reducing organisms to machines.

The discovery of the great ape fueled these debates and Tulpius did not fail to comment on it by adding it to a treatise on human anatomy. It was the very first faithful illustration, not only of a great anthropoid, but also of a chimpanzee. It was in a sitting position:

> Although this is out of the medical field, I will however add to this plot the Indian Satyr which, in our memory, was brought back from Angola and offered to Frederick Henry, Prince of Orange. This Satyr, although a quadruped, presents a human aspect, which is why he is called by the Indians "orang-utan" or "man of the woods". (Tulpius, *Amstelredamensis Observationes Medicae*, 1641, author's translation)

The description remained very general and was limited to the external aspect, completed by an engraving faithful to reality: the feet and hands are those of a great ape, the face and the space between the eyes are those of a chimpanzee; they are not confused with those of the orangutan when they are closer together, and the swollen abdomen can be seen in both of them sitting down, especially since they are older (Fig. 15 in Figure 6.1).

6.1.5. *Edward Tyson (1650–1708), the first dissection of a chimpanzee*

Tyson was born in Bristol. His postgraduate studies took place at Oxford where he became familiar with dissection, and he received his Bachelor of Medicine from Cambridge University in 1676. He moved to London and met Robert Hooke (1635–1703), an eminent 17th century scholar. Hooke was a physicist, he carried out experiments that lent themselves to arithmetic and made it possible to infer the laws of physics. He was a former member of the "Invisible College" which met at Gresham College in London to discuss metaphysics reasoned from geometric mathematics and physics. In 1663, he obtained a charter from King Charles II which granted the college the title of Royal Society (Leguebe 1958). Physiology and comparative anatomy were needed to participate in this research of the laws of physics. This was the case for blood circulation with Harvey, and also had to be the case for breathing, for vision and for locomotion. He therefore encouraged Edward Tyson to continue the work of dissection. He was elected a Fellow of the Royal Society in 1679 at the age of 26 years and obtained an agreement for the corpses of exotic species in St James's Park – the oldest of the Royal Parks in London – to be entrusted to the Royal Society for dissection. Tyson was awarded a Doctor of Medicine degree from Cambridge University in 1680. In 1682, at the age of 32 years, he was appointed prosector of the Royal Society, in other words, made responsible for dissecting animals, and thus became the first anatomist in England.

He was the very first to practice the dissection of a great ape in history, that of a chimpanzee. The exhaustive description was published in English in 1699 and entitled "orang-utan, sive Homo sylvestris: or, the Anatomy of a Pygmy Compared with that of a Monkey, an Ape, and a Man" (Tyson 1699).

Aristotle's anthropoids, or *simia*, now included two new large species, one Asian and one African. Tyson was confronted with the lack of a nomenclature and a reasoned classification. Gessner differentiated between monkeys with tails and those without tails. The two new species were very large, tailless monkeys. The Asian one had an Indonesian name, orang-utan, meaning "wild man of the woods", and written in Latin *sive Homo sylvestris*. This was not a naturalist nomenclature, it was only the Latin translation of the Indonesian name. The genus name *Homo* therefore had no anatomical value.

Tyson named his African specimen "Pygmy". The chimpanzee name appeared later. The very fine illustrations show a young individual in a vertical posture, supposed to move with the help of a cane, probably related to Breydenbach's illustration (Figure 6.2). It was drawn from the front and back before dissection, then emaciated with the muscles and finally with the skeleton in the same postures. Tyson had the Breydenbach–Tulpius-tailed monkey, Gessner's Asian orangutan and Tulpius' African orangutan represented on the same plate (Figure 6.1). Since he published in English, he introduced names that are still in common use: "monkey" for the tailed small simian and "ape" for the tailless great simian. In the French language, this distinction corresponds to *petit singe* and *grand singe*.

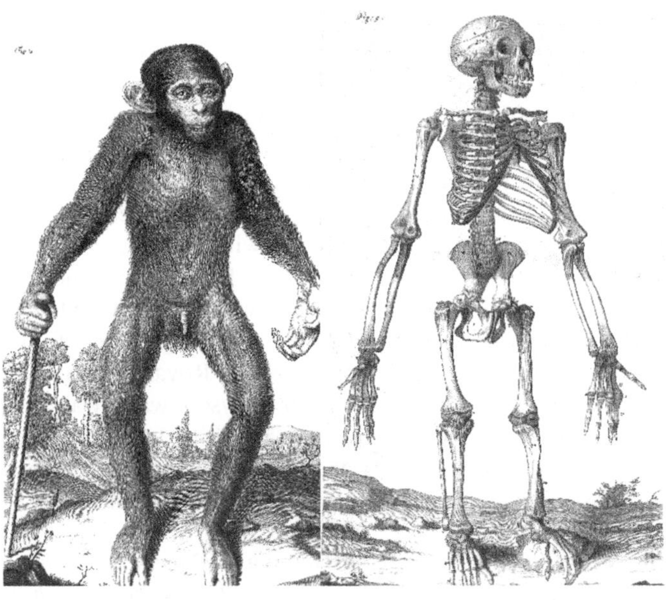

Figure 6.2. *First description of the chimpanzee and its skeleton, Tyson,* Anatomy of a Pygmy, *1699. © UMR 7194 CNRS – Bibliothèque de l'Institut de paléontologie humaine*

So, for the very first time, a great ape was dissected and compared to humans. Tyson described the bones and muscles and cut the brain transversely, trying to highlight what was similar to the monkey and also what was most significantly different from it. This were neither the skeleton nor the teeth, but the brain (cerebrum) and cerebellum (Figure 6.3).

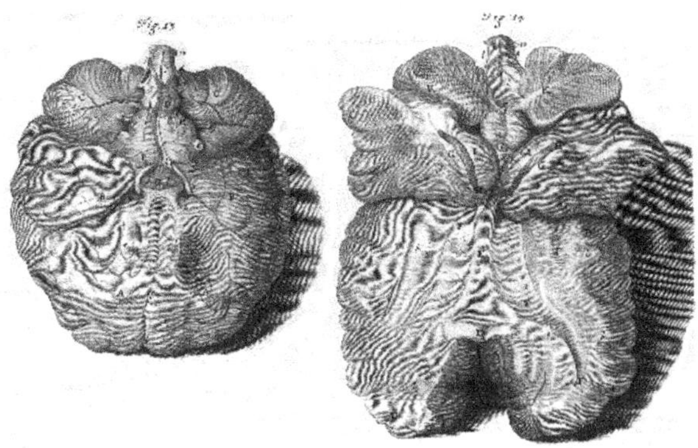

Figure 6.3. *First description and dissection of the brain and cerebellum of a great ape (chimpanzee), Tyson,* Anatomy of a Pygmy, *1699. © UMR 7194 CNRS – Bibliothèque de l'Institut de paléontologie humaine*

The natural history of animals, so gradual in its classification according to Aristotle, showed a considerable gap between the monkey and the human. This gap was beginning to be filled, Tyson's study being exceptional:

> I have made a Comparative Survey of this Animal, with a Monkey, an Ape, and a Man. By viewing the same Parts of all those together, we may *the better observe Nature's Gradation in the Formation of Animal Bodies, and the Transitions made from one to another*; than which, nothing can more conduce to the Attainment of the true Knowledge, both of the Fabrick and Uses of the Parts. (Tyson, *Anatomy of a Pygmy*, 1699)

Tyson then posed crucial problems in the interpretation of osteology by reasoning about both growth and function. He commented on Galen's anatomy by returning to Sylvius and his attitude toward Vesalius. He reminded us that it was essential to understand osteo-muscular and vascular anatomy, taking into account the insertions. According to him, muscle function modified the shape of bones as he observed in children. It was therefore essential to note their location, because although this function differs between humans and other primates, muscular efforts did not modify the skeleton of the Pygmy and humans in the same way.

For Tyson, the great ape was an animal and therefore had no soul (in the sense of a spirit). However, the difference between the brains of apes and humans was shrinking, and what distinguished them was not the erect posture, occasional in apes, but the amount of brain matter. The difference with the chimpanzee (the Pygmy) was too small to justify the spirituality of the human. Thus, Tyson reconsidered his convictions: "organized matter can never produce the soul", which meant that the spirit and its qualities could no longer depend on the structure and quantity of brain matter.

The inner surface of the skull of an ape, and in this case a great ape, was finally observed; the cranial vault had been sawn transversely to remove the brain. It is unfortunate that Tyson did not perform a medial sagittal section to see the differences in straightening. This was in 1699 and no one had yet looked at this primordial difference between humans and anthropoids.

The years went by. The classification of *simia* and humans was still not structured in a systematic comparison of specimens by the same naturalist.

6.1.6. *Carl Linnaeus (1707–1778), the classification of organisms by species and genera*

Carl Linnaeus, or Carl von Linné, was born in Råshult (southern Sweden). He turned to botany at a very early age. He aspired above all to a career as a doctor to escape the career of pastor that his father intended to pass on to him. The college of physicians at the University of Uppsala was not sufficiently constituted to defend a thesis in medicine. Linnaeus tried to obtain his doctorate in the Netherlands in Leiden and eventually obtained it at the University of Harderwijk; the level was less prestigious but it was appreciated for the competence of the doctors who taught there. Unlike Buffon, Linnaeus did not have an academic background and was not familiar with Aristotle's natural history. In 1741, he was awarded the Chair of Medicine at the University of Uppsala and then of Botany. The study of plants required their classification and unity. The reference in botany was John Ray (1627–1705), who set the definition of species according to reproduction: the individuals who belong to the same seed in the lineage of the previous individuals since Creation form together a species. The seed transmits the *eidos*, fixed, immutable, independent of time. Variations are accepted within its limits. Outside these limits, closely related species are united in a genus.

The specialist in the classification of plants by genus was a contemporary of Linnaeus, Joseph Pitton de Tournefort (1656–1708), who held the Chair of Botany at the *Jardin royal des plantes* in Paris until 1706, and then the Chair of Medicine and Botany. Buffon accepted the definition of the species but did not grant a physically constituted reality to the genus. Only the species has a history through its variations linked to that of its environment. Beyond that, similarities simply indicated the distinct lineages present since each was created. The genus was seen as a construction of the mind useful to arrange the species in collections; the genus had no natural history.

Linnaeus's activity was distinguished by a method of binominal classification of plants relying on the reproductive organs, since these are the means of transmitting the seed of their specificity. It integrated the notion of a genus, bringing together very closely related species, and defined a specimen by giving it the name of the genus followed by the species. He extended this principle of classification to animal species after having described new ones sent from abroad that he had to classify for the realization of his *Systema Naturae* which ended up becoming an encyclopedia. After his death, his wife put on sale a large part of his collection (more than 14,000 plants and 3,000 insects) as well as his library and his manuscripts.

The first edition appeared in 1735; the tenth edition of 1758 presented a generalization of its binominal classification. Noting hybridizations between what he considered to be species, he concluded that the genus was the archetype conceived by the Creator; he did not know Aristotle, but it was indeed to his concept of eidos that his idea of the genus corresponded. The Latin translation of the name Man being *Homo*, this word would henceforth qualify human geographical and ethnic diversity.

In 1735 he created the order of *Anthropomorpha*, or "human forms" of the quadruped class (Linnaeus 1735). He created three genera well described by anatomists, including the Sloth (*Bradypus*) and the *Simia*, with five varieties, the tailless monkeys (*Simia cauda carens*), the baboon (*Papio*), the Cercopithecus (*Cercopithecus*), the Cynocephalus (*Cynocephalus*) and the chimpanzee (*Satyrus*). Bontius's orangutan was not known to anatomists and therefore not retained. The genus *Homo* was composed of four varieties, *Europaeus*, *Americanus, Asiaticus* and *Africanus*.

The last edition of 1758 shows a considerably enriched diversity of species. The name "*Anthropomorpha*" was replaced by "Primates", sloths were replaced by bats. A new genus appeared, the primates of Madagascar with the Lemurines forming the genus *Lemur*. The good old genus *Simia* of Aristotle had a large number of new species, especially from America. The *Simia* were subdivided according to the importance of the tail: lengthened, reduced or absent. The chimpanzee and the orangutan were arranged with the tailless monkeys and a newcomer from Southeast Asia that caught attention because of its bipedalism, the gibbon. As for humans, they formed a seperate genus, *Homo*, and a unique species, *sapiens*.

The anatomy of modern humans thus defines both the genus and the species.

6.2. Comparative anatomy at the *Muséum national d'Histoire naturelle*

6.2.1. *Louis-Jean-Marie Daubenton (1716–1799), the occipital hole and the face unified by geometry*

Louis-Jean-Marie Daubenton was a doctor and a cousin of Georges Buffon, who appointed him to be the curator of collections and an anatomy demonstrator at the *Jardin royal des plantes*, in addition to giving lessons at the *Collège de France*. He was also co-editor of the numerous chapters of *L'Histoire naturelle des Animaux*. His meticulous osteological descriptions of current and fossil species, combined with the quality of his reproductions, guaranteed the reliability of his observations. The exceptional number of specimens compared, and even geometrized, allows us to consider him as the founder of comparative anatomy of current and fossil species.

Although Daubenton introduced measurement into comparative anatomy, he did so to classify the fossil bones of the royal collection, which was very progressively enriched. The difficulty of classification was all the more considerable as the sedimentary layers were ancient and comparison was always problematic with current skeletons. Daubenton shared Buffon's opinion about genera, that they were distinguished by clearly visible organizational units with the skeleton. Sooner or later, the naturalist would find himself confronted with an unknown fossil morphology.

In 1740, a French officer on maneuver in New Orleans brought back fossils of some very large herbivores collected in the state of Ohio. They were not attributable to any living genus. In 1749, in his history of the Earth, Buffon admitted that, following a new spontaneous creation, many ancient genera could disappear from the surface of the globe due to a lack of "organic molecules" in the soil. This was not the case for Daubenton, who did not have the same speculative boldness. According to him, since creation was perfect, God provided for the life of all his creatures and no genus could be destroyed.

The idea of the impermanence of species nevertheless made headway. While the "mold" was a divine emanation imposed in Buffon's mind by its adherence to the biblical scale, the acceptance of its geomorphological plasticity constituted considerable progress thanks to the integration of long geological periods in the science of organisms.

Buffon considered that degradation was the rule for the fate of species; they tended toward extinction, just as the Earth tended to cool down. Humans could not be a degenerate form of monkey, so they were an "internal mold" in their own right.

He was convinced that the appearance of humans marked the last phase of his cosmic history, more precisely its completion.

Until now, as we have verified in this long retrospective, no naturalist paid attention to the orientation of the occipital hole, which corresponds to the large orifice that surrounds the segment of the nervous system that extends from the spinal cord to the brain stem located at the base of the skull. Even Linnaeus neglected this cranio-spinal part, which is essential for the connection between the neurocranium and the locomotor skeleton via the spinal column.

On September 1 1764, for the first time since Aristotle's History of Animals, Louis Daubenton presented to the *Académie royale des sciences* an angular comparison of the position of the occipital hole for a dog, a lemur, a chimpanzee and man. The article in French is entitled *Mémoire sur les différences de la situation du grand trou occipital dans l'homme est les animaux* (Daubenton 1764).

The Angola ape is the Kimpézé or chimpanzee of Angola. Daubenton was fortunate enough to have a skull of a young individual, so he compared it to a skull of a young man. According to the illustration, he based his comparison on the dentition, which corresponds to the advanced adolescent stage:

> Man's large occipital hole differs greatly from that of animals in its location; these differences are mainly due to differences in the attitude of the body and the shape of the head. Since man's body and neck are vertical, his head must be balanced on the spinal column, to make all his movements easier, and to keep it on the bony column, which is the point of support given to him by the natural attitude of the human body; therefore, man's large occipital hole is placed approximately in the center of the base of the skull; this hole is not much further from the end (BCE) of the jaws than from the bottom (D) of the occiput: the head is so well placed for its balance, that if one extends the vertical line followed by the body and the neck, it passes through the top of the head. (Daubenton, *Mémoire sur les différences de la situation du grand trou occipital dans l'homme et les animaux*, 1764, author's translation)

The skull was seen from the side. Daubenton's geometric study was exocranial. The angle measured gave the orientation of the major axis of the occipital hole with respect to a straight line drawn between the lowest point of the orbit and the posterior point of the occipital hole. The angle changed from approximately 90° to 0°. The two straight lines forming a compass that closed to such an extent that in humans they merged (Figure 6.4).

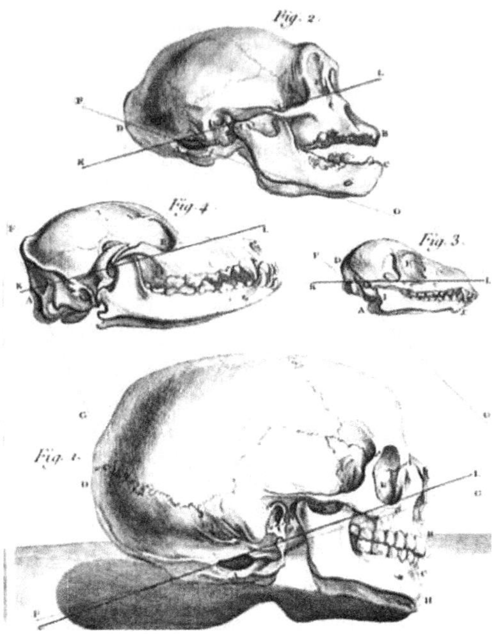

Figure 6.4. *The first comparative geometric study of the orientation of the occipital hole (Daubenton 1764). Source: gallica.bnf.fr/Bibliothèque nationale de France*

Daubenton then drew up an inventory of the differences in proportions, all of which refer to the verticalization of the spine, focusing his attention on the mandible (lower jaw):

> In this attitude, the (human) face is on a vertical line almost parallel to that of the body and neck, therefore the jaws do not extend much further forward than the forehead; they are very short in comparison with those of most animals, because the length of the lower jaw, measured in man from the chin (H) to the posterior edge (I) of the condyloid process, is only half the length of the entire head taken from the chin (H) to the occiput (D), an eighteenth part of the length of the entire body, from the top of the head to the heels; but this last dimension can hardly take place in the comparison of animals with man, because there is no animal whose hind legs have, like those of man, as much length as the body, neck and head taken together and measured from the top of the head to the pubis. (ibid., author's translation)

He noted the proportions of the chimpanzee and correlated the position of the occipital, the orientation of the major axis of the occipital hole and the length of the mandible.

He dealt with the skeleton in its entirety and showed the unity of the geometrical organization plan in relation to the degree of straightening of the spine and thus in relation to the lowering of the occipital hole; the short mandible under the forehead was correlated with a vertical spine and the occipital region lowered toward the vertical axis. He did observe an inverse correlation between the lowering of the occipital pole related to the occipital hole and the retreat of the mandible on the frontal lobe. These were absolutely crucial observations.

Unfortunately, the study remained exocranial; the prolongation of the general orientation of the cervical spine in the center of the skull was not compared. The internal organization of the cranial base concerned by this orientation was not taken into consideration, as if it were only a secondary question that would contribute nothing more than the external appearance might show.

Daubenton did not need to compare endocrania to write:

> The more volume the brain has in comparison with the size of the whole body, the more convex and protruding the occiput is; the further the large occipital hole is from the bottom of the occiput, the more the plane of this opening approaches the horizontal direction; therefore, the large occipital hole of man is not much further from the end of the jaws than from the bottom of the occiput. (ibid., author's translation)

The anatomist did not really give an explanation; he drew up a report and concluded:

> It can be concluded from these observations; 1. that among viviparous quadruped animals, those whose natural attitude and gait is to support and walk with the aid of four feet, have their heads articulated with their necks through the back of their skulls; 2. that this joint is in the middle of the base of the skull of man, because his body has a vertical direction, and that it is carried on two feet only; 3. that animals whose head and neck joints are placed between the middle and the back of the base of the skull are disposed by their conformation to adopt the attitude and gait of other animals and those of man. (ibid., author's translation)

At most, it is possible to understand that, according to Daubenton, it is the vertical direction of the spine that accounts for the position of the occipital hole, itself carried by both feet. There is no functional purpose in this text, such as "man is made to walk"; the vertical posture is related to metaphysical considerations that escape reason because "nature executes the laws of the Supreme Being".

Daubenton left the internal base of the skull in the white of the map, giving more importance to the orientation of a void, the occipital hole, than to the orientation of the bones and their limits which make a hole show such or such orientation. He overlooked the internal study of the base, which would have allowed him to see which bones confer this orientation by their orientation and by the limits of their edges, how together they form the periphery of the orifice. Indeed, there is also a "compass" that closes between the face and the base of the skull, but the skull would have had to be sawn in half in the median sagittal (vertical) plane as in Leonardo da Vinci's drawings to show this. Probably the rarity of monkey skulls and especially that of the chimpanzee explains why this was not done.

6.2.2. *Georges Buffon publishes his own "Nomenclature of Apes"*

Buffon had the opportunity to see a chimpanzee in 1738, one year before he took up his post at the *Jardin des plantes*. Daubenton gave a good osteological description that made it possible to distinguish them from humans; his master was already writing in 1749:

> We see man whose background is all the same slowly modified by the continuous action of climate, soil, habits. (Buffon, *Histoire naturelle de l'Homme*, 1749, author's translation)

Georges Buffon published volume XIV of the *Histoire Naturelle des Quadrupèdes* in 1766. His nomenclature of primates is fastidious and speculative; it expanded on the difference between quadruped and quadrumane, which had front feet which were hands with a palm and hind feet which looked like hands. The four extremities allowed movement of the body in trees by gripping, with the strength of the arms and feet. He insisted that humans were bipedal and biman (two-handed). The differences were radical, at the very root of the architectural "conception" of the body. Buffon had great admiration for Isaac Newton and reasoned in terms of forces of attraction, on the atomic scale as well as on the terrestrial and cosmic scales. He was not mistaken: the axial skeleton of the great ape, such as the orangutan, was no longer subject to the force of terrestrial attraction; it was suspended, moved by arms considerably longer than legs which did not have to support the weight of the body, whereas humans had two feet made to support the axial skeleton and hands which have no locomotor function.

"Man is the only one that is both biman and bipedal, because he is the only one who has two hands and two feet"; he defined the order of bimanes for humans, and quadrumanes to include apes.

The difference in the straightening of the skull base had still not been observed, the internal surface of the neurocranium and the zone of passage from the brain stem to the spinal cord was of no interest to anyone. The orientation of the cerebellum was not evoked, even though its role in the control of balance had been recognized by physicians since Antiquity, with Erasistrate of Ceos. The concept of psychomotor balance was lacking, as in the process of learning to walk, which involves the development of the brain (cerebrum), cerebellum and the sense organs.

In 1778, Daubenton succeeded in transforming the famous medical chair of the *Collège de France* into a Chair for Natural History. Buffon died 10 years later, in 1788. His faithful assistant succeeded him and became the new Superintendent of the *Jardin royal des plantes*.

The choice of characters to evaluate the proximity of the ape to the human remained arbitrary. The emphasis was on locomotion and behavior, not on the development of the nervous system and the membranous and skeletal envelope that protects it. Thus, a representation of a monkey standing with a stick humanized it, whereas, in the wild, the monkey moves most often in quadruped mode if arboricolism (tree-dwelling) is not its preferred mode. Moreover, comparative anatomy, which is most often supported by dissection, was almost non-existent.

Borelli showed how the living body can be the subject of a geometrical mathematical study, both for the comparison of statics and kinetics. René Descartes made it a machine, a mechanics, then, thanks to Isaac Newton, Gottfried Leibniz made it an organism animated by its own internal forces.

There was so much still to be understood, the great unknown of the organs being the nervous system. The differences between humans and apes are related to cognition and language. Apes are deprived of articulated language and this deprivation of speech calls for attention: is it neurological, functional, architectural or physiological? Should we distinguish these parameters or make a system out of them?

The question took a new turn with the first dissection of an orangutan at the end of the 18th century.

6.2.3. *Petrus Camper (1722–1789), the first dissection of an orangutan: the ape does not speak*

Petrus Camper was born in the Netherlands. He obtained his medical degree at the age of 24 years. He was soon appointed Professor of Anatomy and Surgery at the Franeker Chair, then in Amsterdam and finally in Groningen. The opportunity arose to dissect a young orangutan at the stage of temporary dentition (milk or deciduous teeth). His study, published posthumously in 1791, was a landmark in many ways (Camper 1791).

Camper was socially well-placed to observe several individuals; he dissected three of them and observed at least four others; some individuals belonged to private individuals, others to the menagerie and the collection of the Stathouder of Amsterdam (the General of the United Provinces). He demonstrated that this Asian great primate had nothing to justify the name *Homo sylvestris*, "Man of the Woods"; its skeleton had no morphological character in common with that of humans.

Its vocal tract diverged from that of humans through an architecture which was not verticalized and a hyperspecialization of the larynx, with a membrane fixed on the hyoid bone, the small free bone located just under the mandible. This membrane formed a laryngeal sac that inflated with air and allowed the emission of powerful sounds. For the Dutch doctor, it was therefore obvious that the orangutan was mechanically and physiologically incapable of articulating and uttering words. Articulated language was considered to be the expression of thoughtful intelligence, reason and possession of a soul. Until then, geometrical study had made it possible to infer physical, mechanical and dynamic laws regulating the morphology of materials such as bones.

Camper applied, for the first time, a protocol of angular measurements in order to compare the projection of the face of the great ape, unable to speak, with that of Homo sapiens. *He sought to establish correlations between this angle and the intellect while developing a physiognomic approach to facial expressions.*

His method consisted of placing the skull in profile in a geometric marker centered on the ear hole (the bony ear) and aligning it on a straight line passing through this point and the anterior end of the nasal orifice. The skull fitted into a space with four sides, each corresponding to the tangent passing through the outermost points: the top and back corresponded to the cerebral part, with the top of the skull and its posterior part, while the front and bottom corresponded to the facial part (Figure 6.5).

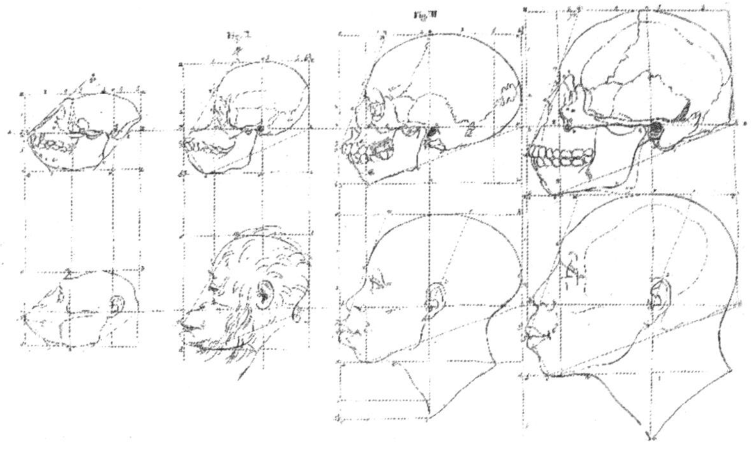

Figure 6.5. *Angular geometry of the relationships between the upper and middle face and the lower face, Camper,* Naturgeschichte des Orang-Utang, *1791.*
© UMR 7194 CNRS – Bibliothèque de l'Institut de paléontologie humaine

Then, he drew straight lines that joined homologous points between species, as shown in Figure 6.3, and measured the angle of prognathism. Camper applied this protocol to the adult skull of a small ape, a young orangutan, and adult human skulls of different ethnic origins. He also applied it to the growth of humans, from the birth stage to the toothless old age. The angle of prognathism was more closed in a small ape (42°), it opened in the great ape (58°), and in humans it was even more open (70° to 80°) since the face is vertical. As for any species whose population is high and scattered around the planet, human diversity was substantial. This study was not based on statistics; it compared the stage of the temporary teeth of the great ape with the stage of the adult teeth in humans, which was not significant, because bone tissue follows reverse growth trajectories for apes and humans.

6.2.4. *The premises of a gradualist and racial anthropology*

Camper then built a racial anthropology based solely on this angle, inevitably graduated since it was quantified. When this monograph was published, Buffon had been dead for three years. However, this distribution of skulls according to an angular value scale was meaningless with respect to two fundamentals: 1) *H. sapiens* had one and the same "internal mold" according to Buffon, with a geographical diversity that depended on climate and living conditions as for any other animal species; 2) there was no gradualism in natural history between the great apes and *H. sapiens*.

By establishing a gradualism between the prognathism of the ape and the variety of prognathisms of *H. sapiens*, called racial, Camper thus committed a break with this system. His graduation was arbitrary and had no natural foundation.

However, he classified human diversity according to this graduated angular scale that he intended to correlate with lifestyles and civilizations. According to Camper, the statues of ancient Greek figures represented the most balanced profile (90° angle), that of the most accomplished civilization from which the European culture of the 18th century derived. He sought to demonstrate correlations between prognathism and intelligence by positing the postulate that a prognathic face was indicative of an underdeveloped intelligence.

The internal base of the skull was always left out; Leonardo da Vinci's geometrical study of the sagittal section was little known, still dormant in collections scattered throughout Europe.

Camper thus left aside the physical relationship between the articulation of the mandible and the part of the base of the skull that straightened, although it was only "articulation" that communicated the effects of the verticalization of the nervous system to the face, while the encephalic territory of straightening was the brain stem, which had no cognitive functions.

7

The Transition from the 18th to the 19th Century: Birth of Paleontology and Comparative Anatomy

By introducing the very long geological periods and tectonic revolutions of the globe, Buffon made it possible to conceive of the extinction of genera and their replacement by new and better adapted molds. The study of the position of fossils in the soil, known as oryctography, is still the work of great and enlightened collectors with rich libraries.

7.1. Oryctography or the study of the disposition of minerals and fossils in the soil

7.1.1. *François-Xavier de Burtin (1743–1818), a leading European collector*

François-Xavier de Burtin was a physician. In 1784, he was a member of the very young *Académie des sciences et Belles-Lettres de Bruxelle*s which published his work *L'oryctographie de Bruxelles* (Burtin 1784). He was a first rank collector: his cabinet of natural curiosities contained no less than 12,000 geological, mineralogical and oryctological pieces with 5,000 "petrifications", including 4,000 fossils of marine invertebrates (shells, sea urchins, starfish, corals and madrepores) and seeds of tropical species. These fossils were collected in the region of Brussels and, as much as possible, in their stratigraphic context. The book is beautifully illustrated with 32 engraved and colored watercolor plates. Burtin developed a personal reflection on these fossils, recognizing the climate changes and the great antiquity of the globe, as well as the death of species; it was no longer realistic to believe the contrary. He also added a paragraph on an object that he considered legitimate to classify with the "fossils" or "monuments", since it was made of stone and collected at depth. It was a "stone axe" collected *in situ* in a layer of fossiliferous deposits containing fossilized coconuts and under two other layers containing turtles, oysters and nautilus (Plate XIII, natural size, Burtin 1784).

He commented at length on this discovery, which he considered to be, rightly, invaluable:

> Although the stone axes, which owe their form only to the hand of men, seem at first sight not to be among the objects relating to the aardvark, their stony substance, the places where they are found, and above all their singular relationship with the physical and moral history of our globe, rightly assign them a place here. Wallerius arranges them under artificially figured stones, which he calls *Lithoglyphi figurati arte facti*, and, since Greek is needed everywhere, he could have called them technoglyphites, which expresses in a flash an *artificial figured stone* (…). The one I illustrate, of natural size, is the only one found in our surroundings, *it seemed to me such a respectable monument of the remotest antiquity*, that no one that looks at it can be indifferent, nor treated so lightly. It was found in the quarry of the mill at Loo (…). This situation under three layers of such petrifications leads to such singular consequences, in a word *it says so much to the man who thinks*, that I thought I had to take the most meticulous precautions to verify it. Fortunately, the three workers who were working in this quarry at the time of the discovery are still alive and ready to testify to all those who wish (…). (Burtin, *Oryctographie de Bruxelles*, 1784, author's translation)

Burtin was not able to situate the presence of this human activity in a scale of the history of the globe, and it was to cosmologists that he turned:

> I will refrain from this rare observation, the inferences it makes about the age of the globe, and its various eras … but I am happy to have been able to provide cosmologists with a very proven fact, which they can make good use of. (ibid., author's translation)

7.1.2. *The French Revolution: naturalist audacity faced with the fury of the Terror*[1] *(1792–1794)*

This magnificent collection suffered following the French Revolution of 1789. The first Republic was declared in 1792. The *Jardin royal des plantes* became the

1 A period of the French Revolution which saw many massacres and public executions.

Muséum national d'Histoire naturelle by a decree of the National Convention dated June 10 1793, which changed its statutes. Daubenton was elected director by his peers and professors. Twelve new chairs were created: Anatomy of Animals, Human Anatomy, Botany in the Countryside, Botany in the Museum, Culture (agriculture, gardens, etc.), General Chemistry, Geology, Mineralogy, Vertebrate Zoology and Invertebrate Zoology. A chair was especially dedicated to Iconography while "The Chemical Arts" disappeared. Daubenton was appointed Chair of Mineralogy.

France occupied the Kingdom of Belgium from 1792 to 1793, and the Convention appointed a commissioner to recover the most remarkable pieces of Burtin's collections, 600 specimens sent to the *Jardin des plantes*. Daubenton necessarily witnessed this event, which represented a real war theft. Following a trial, Burtin obtained restitution (Groessens 2008).

Discreet, an admiring observer of nature, endowed with great curiosity and a concern for the transmission of knowledge, Daubenton avoided conflicts. He lived through the chronic agony of the "Terror" from 1792 to 1794. Those in charge, ever more directive and invasive, imposed lines of research on the scientists of the *Académie des sciences*, such as the improvement of war machines. The members refused to put the values of science, intended to serve life and peace, to the service of death and violence between populations. In retaliation, the *Académie des sciences* was dissolved in 1793. Many members made themselves scarce in order to avoid the revolutionary courts. One notable member, Claude-Louis Berthollet (1748–1822), maintained relations with the disastrous and fearsome "Committee of Public Salvation" to defend the rights of science.

The *Jardin royal des plantes* avoided being broken up. Buffon's institutional work owes its survival to Daubenton's discreet temperament and his work on sheep for the common good, as well as his popularity. As Georges Cuvier reminds us, his research, useful in improving living conditions, saved him from "the most ignorant part of the people", from "the fury", and from mass executions (shootings, drowning and guillotine):

> In 1793, by a reversal of ideas, the most ignorant portion of the people had to pronounce on the fate of the most educated and the most generous: the octogenarian Daubenton needed to ask an assembly called the section of the Sans-Culotte for a paper whose equally extraordinary name was a *certificate of citizenship*. He had acquired through this work a kind of popular reputation which was very useful in dangerous circumstances (…). What he did for the improvement of our wool will forever deserve the recognition of the State, to which he gave a new source of prosperity (…), training educated shepherds, propagating the practice of his method (…). A professor, an

> academician, would have had difficulty in obtaining it: a few sensible people, who mingled with the furies in the hope of containing them, presented him with the title of shepherd, and it was the shepherd Daubenton who obtained the necessary certificate for the director of the *Muséum national d'Histoire naturelle*. This document exists: it will be a useful document, although less for the life of Daubenton than for the history of that disastrous period. (Cuvier, *Éloge historique de Daubenton*, 1861, author's translation)

Buffon's only son was destined to succeed his father. He embraced a career as an officer and later married one of Daubenton's nieces. He died by guillotine on July 10 1794. Robespierre, who established the Terror, was guillotined in his turn on July 28 1794. In 1795, the Convention adopted a new constitution which allowed the creation of the *Institut national des sciences et des arts*, a substitute for the *Académie des sciences*. The "shepherd" Daubenton, director of the nation's Natural History Museum, was appointed resident member of the anatomy and zoology section. The regime led France into economic disaster; a *coup d'état* led by a young general, Napoleon Bonaparte, put an end to it on November 9 1799. Daubenton died on December 31 1799 with his century and with the bloody burial of the French monarchy.

At the age of 83 years, Daubenton passed the reins to a young generation of naturalists who were sensitive to the observation of living beings thanks to the increasing number of works in botany and zoology. However, this interest was seen as a secondary activity by the Encyclopedists (Diderot & d'Alembert 1762). Balan (1979, p. 138) evoked contempt for them, as classifying current and fossil species by comparative anatomy and biometry "was useless". Life was absent from the classifications, the nomenclators were not zoologists, the ordering was only a great construction game by interlocking "Russian dolls" that revealed nothing of the laws or principles that constructed organisms and ordered them in relation to each other. Energy was not their preoccupation; it was a bit like the preoccupation of zoologists studying physiology. It became a priority for the development of the economy and victories on the battlefields.

The key work on machines and mechanisms was Diderot and d'Alembert's *Encyclopédie des sciences, des arts et des métiers*, consisting of 17 volumes of texts and 11 volumes of plates published between 1751 and 1772. The forces described were animal, human, hydraulic and aerodynamic, but not yet those of the steam boilers. In France, these were used to dry grain or refine sugar, but not as a driving force as shown in the illustrations and descriptions of the Encyclopedia in the volume dedicated to agriculture and economy. France was lagging behind; in England, two engineers had developed steam engines to improve production efficiency. James Watt filed the first patent for a machine in 1771 and Richard

Arkwright inaugurated the first mechanized cotton spinning mill. This was the beginning of the Industrial Revolution.

Wars encouraged monarchs to train military engineers to improve ballistics. This science studied the movements of bodies launched into space and thus tackled head-on the fight against gravitation and the problems of the quantity of movement to be imparted to projectiles. The devices needed to be optimized at all points: the ease of movement, the rate of fire, the distance and accuracy of the shot, the power and thus the energy to be supplied to the projectiles were subjects of study.

7.1.3. *The premises of the Industrial Revolution: energy and thermodynamics*

At the beginning of the 19th century, applied mathematics was essentially dedicated to mechanics. The properties of matter at the molecular scale, with the attractions between atoms, the basis of chemistry, electricity, magnetism and heat, were not yet at the stage of mathematical formalization. What could be measured and what equivalence equations could be discovered?

Nicolas Sadi Carnot (1796–1832) was the son of general and revolutionary Lazare Carnot (1753–1823), one of the founding fathers of the French Republic in 1792. He voted for the death of King Louis XVI in 1793 and was Napoleon Bonaparte's War Minister in 1800, before resigning. Lazare Carnot was a mathematician, military engineer and the author of the *Essai sur les machines en général* (1783) in which he addressed the problem of the loss of the quantity of motion imparted to a machine, when this motion ceased. He did not evoke the driving force of heat. Member of the Physical and Mathematical Sciences Class of the *Institut national des sciences et des arts*, he personally took care of the scientific education of his two sons, Sadi and Hippolyte. The latter became Minister of Education in 1848 with the advent of the Third Republic. Sadi Carnot, on the other hand, had no political or military ambitions. He was a spirit versed in research and focused on the most fundamental mathematical question and the most complex to mathematicize: the permanence of energy. As long as this question was not thought out in depth, it was obvious that the origin of the first cell cycle, i.e. the origin of life and its propagation, remained repressed in the darkest areas of intelligence, with the only alternative being productions of the mind as fabulous as Buffon's "internal mold". The son Carnot studied at the two *Grandes Écoles* that produced the elite of military engineers, the new *École Polytechnique de Paris* and then the *École de Metz*.

In 1800, Bonaparte granted a state pension to a certain Nicolas-Joseph Gugnot (1725–1804), a military engineer known since 1769 for his steam car project that had interested King Louis XV. The machine worked in principle, but the boiler was not adapted to the mechanics of the chassis. It is very likely that Sadi Carnot was aware of it. Steam engines fascinated him. He wondered about the operation of a heat engine, following an unprecedented intellectual path, starting from the motion power of fire. He died of cholera at the age of 36 years and produced only one work written at the age of 27 years and published in 1824: *Réflexions sur la puissance motrice du feu et sur les machines propres à développer cette puissance*. Carnot's intellectual depth gave this work the founding scope of a new physics which would take the name of thermodynamics.

The primordial physical reality that interested him was the perpetuation of energy production to maintain a movement, and its optimization to limit inevitable losses of energy. Any movement exerts friction, whatever the state: gaseous, liquid, viscous or solid. This is why the system must be supplied with forces to fight against the Earth's attraction by moving its mass (inertia) and also against resistance in the opposite direction to the direction of movement, due to contact with a different "body". The difficulty for the machine invented by humans, or for any organism trying to keep itself alive, is the regular supply of energy and components for its construction due to wear and tear. The difference between the machine and the organism is nevertheless radical because the machine does not have the memory of its construction, nor the memory of how it works in order to maintain itself, and it does not reproduce itself along with instructions for use. Or, more precisely, the instructions for use with the stages of construction, of all the materials, their shapes, their position and their articulations, do not reproduce the machine along with the way it was designed. The first cell cycle, on the other hand, was endowed with this prodigious phenomenon that we call "life", which has not stopped reproducing since.

For a steam engine, if the fire is no longer fed, the water temperature drops, the steam disappears and movement stops; the whole is in equilibrium, everything is static, nothing moves, everything is cooled. The fire and the water supply must therefore be constantly maintained so that movement is perpetuated. For the organism, it dies if it does not feed; for the species, it is the dissipation of the memory of construction during generation that provokes abortion; embryogenesis does not develop.

To attraction, which is a force between infinitesimal entities (waves or corpuscles), is added thermodynamics, which is the energy or degree of agitation of these entities. The study of water vapor allowed Sadi Carnot to follow this reasoning: the more water particles are agitated (in other words, H_2O molecules), the more the collective movement is extended and "hot", exerting pressure on a surface like a piston. This piston can be articulated to a wheel by a set of drive shafts and an axis of rotation.

Three new principles were stated; they concerned any process requiring energy to maintain itself in its dynamics, that is to say, in a cycle of reproduction.

7.1.3.1. *Life, between extinction and evolution, a fourth thermodynamic principle*

A cell that reproduces is a cyclic dynamic system. The first principle of thermodynamics is conservation of the energy necessary to maintain the equilibrium of the system, i.e. its reproduction in the same state of equilibrium and organization. The system maintains its equilibrium if the internal dynamics receive enough matter and potential energy. The system is said to be "open" because it exchanges with its environment: the energy consumed and dissipated must be replaced, the quantity of molecules or ions is stable. The dynamic trajectory is cyclic: there is nothing new.

The second thermodynamic principle, known as "Carnot's principle", is derived from this: if the system is physically closed, the cell cycle depletes energy and the dynamic ceases, there is no more animation, the cells degrade and the organism reaches a state of static equilibrium. Duration becomes an arrow of time, which is observed by the passage from an unstable equilibrium to a state of greater stability, that is, the disappearance of internal movements and dynamics, that is, death, followed by disorganization. The production of heat due to organized cellular work is called "entropy".

The third principle follows from this: entropy is zero when the temperature is absolute zero, nothing moves, which is in fact unrealistic since everything is in motion (Descartes) or dynamic whatever the scale (Leibniz).

The idea of the evolution of species through a change in the mechanisms of generation emerged at the end of the 18th century, but the science of thermodynamics was still too "immature" in the 19th century for naturalists to understand that it provided the principles to begin clarifying the problem of the natural origins of life, or the first cell cycles. However, as long as the process of emergence of the first cell cycle was not formalized by laws or principles, it was vain to imagine that the complexification of unicellular organisms could be formalized. It was necessary to begin by highlighting the laws of construction with the only possible method: mathematics.

A cycle is a construction manual that reproduces itself; it is a sort of living memory of its own construction.

As long as the cycle reproduces itself by compensating for its energy losses, it is in a state of stable dynamic equilibrium. When the cycle becomes more complex by no longer reproducing the cycle but a sequence of cell cycles that diverge in their specialization, the system follows an irreversible time arrow that is not death, but exactly the opposite. It is an even longer time arrow for the existence of the organism, a prolonged duration: it is ontogenesis.

From the first unicellular cycles to mammals, it was thus an arrow of internal time that had irreversibly propagated itself and the opposite of that which led to inertia, death or degradation. This would take the name of negentropy, the "arrow of life".

Instructions for use become more complex while maintaining this property of transmitting principles. It was a "fourth principle of thermodynamics", the transmission of a unicellular or pluricellular memory of organization, namely, embryogenesis; the problem of the transmission of changes by the paths of generation would remain insoluble.

This was the great difficulty, but the formidable challenge, that was emerging with the nascent 19th century.

7.2. Georges Cuvier (1769–1832), the French Revolution and the revolution of the globe

7.2.1. *The natural sciences at the heart of the "Terror"*

Georges Cuvier attempted to bring order to the chaos of a naturalistic thought that struggled to structure itself, to synthesize itself and to impose itself as an autonomous access to knowledge that did not have to account for a coherence or an inadequacy with biblical doctrine.

Georges Cuvier was born in Montbéliard, which was then part of the Duchy of Württemberg (Germany). His interest in the natural sciences began in his adolescence. Higher education was expensive and his father could not afford the schooling reserved for the wealthy elite. He was given the opportunity to study at the Caroline University in Stuttgart, the Ducal Academy of Württemberg, which had a faculty of medicine, law and administrative and military sciences. The subjects taught offered a broad and solid general education. For example, Cuvier read Buffon's *Natural History* and learned about field research with fellow students. The Count d'Héricy noticed him. He hired him in 1788 as a tutor. Established in the Caen region, Cuvier devoted his free time to collecting fossils and other bones gathered in the wild and to dissecting what fell into his hands (a parrot and the countess's cat, mollusks and fish). The "Caen stone" was rich in very large mineralized bones, which later proved to belong to dinosaurs.

If Cuvier found bones of ancient carrion at some distance from each other, and if he tried to classify them, he did not put a mouse femur with a pigeon humerus. He understood that the skeleton was a coherent architectural structure, built according to physiological and mechanical logics that he would later call "principles of correlations" and "subordinations between organs and bones". This skeleton reconstitution from scattered bones, present or fossilized, was impossible without a reference collection of present species.

As the Revolution gained momentum, the Count d'Héricy's family left Caen for a hamlet located north of the mouth of the Seine, hidden in deep countryside, and lived there confined during the time of the "Terror" (Beaurepaire 1866). Cuvier remained at his service while exercising the patriotic and republican function of secretary/registrar of the small commune. In times of great scarcity, it was advisable not to waste his time on useless occupations such as the collection of actual and fossilized shells. To serve knowledge for knowledge's sake was a luxury that the Republic had to preserve by saving the *Académie des sciences*. In 1794, a form of resistance was organized against the ignorance of the "furies" who dissolved it. He managed to share his thoughts with the chief physician of the hospital in Fécamp, Henri-Alexandre Tessier (1741–1837), a member of the *Académie des sciences* since 1783, and a correspondence was established with Étienne Geoffroy Saint-Hilaire (1772–1844), a young professor freshly appointed by Daubenton to the Museum's new Chair of Vertebrate Zoology created in June 1793. In the most troubled times of the "Terror", Cuvier chose to keep his duties "until the fury of the revolutionaries against the ecclesiastics subsided" (ibid.).

Robespierre was condemned to death and in turn guillotined. Cuvier resigned from his position in February 1795 and went to Paris. The young man was gifted with a capacity for analysis and synthesis unusual for his age. Tessier recommended him to the new *Institut des Sciences* and the new Professor of the Museum, highlighting his qualities as an anatomist, a man of the field and also a draftsman, a quality necessary to understand shapes and pursue their geometric analysis. Louis Daubenton joined him in 1796 as deputy to the Chair of "Anatomy of Animals".

Following Daubenton's death, Cuvier took over his Chair of Natural History at the *Collège de France*, while Étienne Geoffroy Saint-Hilaire, holder of the Chair of Zoology, was elected Director of the Museum. The two men shared their reflections exchanged since 1795.

When Cuvier took up his duties, fossils were still geochronological markers and therefore preserved in geological collections in Earth Sciences. He contravened this relegation by recognizing them as ancient species to be studied as the remains of living organisms. He changed the name of his chair to "Comparative Anatomy" and engaged in a vast program of natural history of extinct species. The comparison of fossils with present-day skeletons complicated the classification because of the lack of comparable living species and required a reconsideration of the criteria. The reflection was therefore refocused on the skeletal framework. While Cuvier wanted to rethink the classification that integrated the "lost species", it was necessary to series the different levels of organization from the general to the particular and to give himself a method, including the duration of ontogeny and the duration of large geological scales. He needed to demonstrate an ability to synthesize knowledge, common sense and discernment, with a height of vision capable of embracing complexity at all scales.

If Cuvier wanted to classify the skeletons of fossil and present-day vertebrates in all rigor and according to a guiding thread, he would have had to compare all the trajectories of tissue morphogenesis along the cranio-caudal axis since embryogenesis. This guiding thread had been intellectually present since the Renaissance, but scattered over time and the geographical space of Europe, as a hive in which the structure of knowledge was built over the centuries, a movement of the mind shared by individuals who passed it on to each other with their learned societies and from generation to generation by the universities, with more or less success depending on their regents. These truth-loving naturalists were searching for an understanding of what was seed, fertilization or generation in terms of embryogenesis in the broadest sense (development from embryo to full-term fetus); they wanted to understand infant development and growth, biomechanics modeled from head to toe, the physiology of blood circulation and the physiology of pulmonary respiration. The least well-understood organ was the nervous system, all the more difficult to observe in its organization and its operation because it complexifies between that of a snail and a frog endowed with an encephalon; it becomes more complex in a parrot, and even more so in a sheep, then in a monkey and human.

Fossil invertebrates were collected in the Alps: seas had disappeared, ocean bottoms had collapsed, sediments had accumulated, then had been folded, compressed and raised. Everything was being put in place so that the idea of a natural transformation of organisms to the rhythm of the tectonics of the cooled globe could be explained. The idea of evolution could now sediment in a few daring minds, even though there was no physicist or a micro-anatomist able to tell how the microscopic world of seeds could become the epi-phenomenological sponge of ecological changes.

No matter if the deepest roots in time, and the most minute ones in space, in other words the molecular organization of the seeds of fossil species, were lost forever, the step would have to be taken. It would not matter that two sex cells that combine their inheritance must first respond to the mathematical and physical laws of universal attraction before they could give birth to a cranio-caudal reconstruction. This scientific rigor was mathematically impossible in the 19th century. However, without this condition, which is the basis of a true scientific theory, there was no possible perspective toward a true scientific discovery, i.e. the mathematical writing of an unknown physical law. This rigor was formulated by a brilliant young mind at the University of Stuttgart, Karl Kielmeyer (1765–1844), an educator and then a friend of Cuvier's from 1784 to 1788.

7.2.2. *Karl Kielmeyer, Georges Cuvier's great comrade and gifted youngster*

Karl Kielmeyer, born in 1765 in the Duchy of Württemberg, was a lecturer in Natural History at the Caroline University in Stuttgart when Georges Cuvier came to Stuttgart to pursue his higher education from 1784 to 1788. Kielmeyer was four years Cuvier's senior and the two men befriended each other and continued to exchange their thoughts through letters. In 1788, when Cuvier left Germany for Normandy, Kielmeyer was appointed at the age of 23 years to head the Stuttgart Museum of Natural History. In 1792, he was Professor attached to the Chair of Chemistry at the Faculty of Medicine. The young man was able to draw the main guidelines from the knowledge of his peers whom he taught. In 1793, he presented a synthesis of his reflections that impressed the academic audience. Contrary to some interpretations of his text (published in 1995), Kielmeyer never studied a human embryo, it was materially impossible. Further, he would never have been able to say that human embryogenesis recapitulated the major stages of past typogenesis, since no fossil primate was known in 1793, while the classification of existing species was still rudimentary. The text was short, about 20 pages long, and explained the main points of the scientific approach. His knowledge was multidisciplinary, or encyclopedic for the time, and he understood how, with measurements and a good knowledge of arithmetic, it was possible to infer the laws governing the organization and animation of everything that exists as a closed system. His concepts were published under the title: *Über die Verhältnisse der organischen Kräfte unter einander in der Reihe der verschiedenen Organisationen, die Gesetze und Folgen dieser Verhältnisse* ("On the Relationships of Organic Forces among Themselves in the Series of Organized Beings, and on the Laws and Consequences of these Relationships") (Kielmeyer 1995). He shared his concepts with Cuvier:

> The theorist's ambition is to reduce all vital phenomena to a few simple laws whose formulation imitates the rigor of mathematics. The aim is to define the forces at work in living nature, in its totality, to quantify as far as possible their positive and negative effects in the interaction of phenomena, and thus to arrive at the determination of the laws of life. (Cuvier, in Taquet, *La naissance d'un génie*, 2006, author's translation)

Was he an evolutionist?

The term "vitalist" attributed to him simply designated the posture of a scientist who knew how to distinguish between dynamic processes of self-replication, specific to living things, and those that governed the growth of minerals, which do not reproduce themselves. It was not a question of vital breath, but of the properties that formed life itself. This difference was evoked with Lamarck: a crystal was not a mode of construction that was, at the same time, the mode of use of its construction. It was not a cycle, or a dynamic loop whose conceptual difficulty still persists in the 21st century.

Karl Kielmeyer did not push his reasoning to the limits of the boundaries between the mineral and the living because understanding the simple laws that allowed the first cell cycle to emerge was the same as understanding how the building instructions were memorized to reproduce themselves. When stated in this way, the evolution of life, by iteration and integration of its major stages structuring a new embryogenesis, could be understood more clearly, but abstract representation was inaccessible to our linear way of thinking.

7.2.3. *Étienne Geoffroy Saint-Hilaire, the unity of the animal composition plan*

Geoffroy Saint-Hilaire (1772–1844) was born in Etampes, south of Paris. He was interested in the natural sciences from an early age and attended the classes of Louis Jean-Marie Daubenton at the *Collège de France* and those of Antoine-François Fourcroy (1755–1809) in chemistry at the *Jardin des plantes*. He then met Bernardin de Saint-Pierre (1737–1814), an ecologist before his time, who was appointed Intendant of the *Jardin des plantes* in 1792. He wrote a report entitled *Mémoire sur la nécessité de joindre une ménagerie au Jardin national des plantes* and probably considered adding the young Geoffroy to it. He supported him along with Daubenton for the position of under-keeper and assistant demonstrator in the Cabinet of Natural History. He obtained this position for him in 1793. A few months later, Geoffroy was elected Professor to the Chair of Zoology at the new National Museum and found himself in charge of his future menagerie. It was during the heart of the Terror in 1794 that Georges Cuvier entered into correspondence with Geoffroy.

The two men thus found themselves in 1795 discussing their interpretation of the homologous or analogous characteristics of the organs and structures of invertebrates and vertebrates. This discussion touched on the very foundations of the notion of the organization plan and was at the origin of their division, closely followed by Goethe, the great German poet (Johann Wolfgang von Goethe 1749–1832).

Geoffroy published his first study of a primate species, the Maki of Madagascar, (the lemurid *Lemur catta*) in 1795. It was from this study that he defined his notion of the unity of the compositional plan of living beings, or the unity of organization:

> A constant truth for man, who has observed a great number of the productions of the globe, is that there exists between all their parts a great harmony and necessary relationships; that it seems that nature has enclosed itself within certain limits, and *has formed all living beings only on a single plan, essentially the same in principle, but has varied in a thousand ways in all its accessory parts*. (Geoffroy Saint-Hilaire, *Mémoire sur les rapports naturels des Makis Lémur*, 1795, author's translation)

7.2.4. *The laws of animal oeconomy*

Cuvier and Geoffroy worked together. In 1794, the revolutionary army defeated the troops of the Stathouder of Amsterdam. The Convention claimed the collection of the Cabinet of Natural History and had it transferred to the Museum in Paris in October 1795. Although this one contained a skeleton of an orangutan at the stage of temporary teeth known to Camper, they had never seen other specimens.

The two anatomists therefore applied Camper's facial angle method by modifying it and published their study under the title *Histoire naturelle des orang-outans* (Geoffroy Saint-Hilaire & Cuvier 1795) and attributed the skeleton to a Mandrille (Cercopithecidae).

This study changed to mockery because it was in fact an orangutan at adult stage, very far from the infantile form described in the few articles on the orangutan. Cuvier made amends by acknowledging that he did not take growth into account. In the end, the facial angle was abandoned because it did not provide any taxonomic information, especially since it changed with the development of the teeth. The two anatomists paid no attention to the internal organization of the skull and the orientation of the bones determining the position of the occipital hole and thus of the spine.

Cuvier shared his thoughts with Geoffroy Saint-Hilaire on the principles of a new classification that could integrate fossils and the "laws of animal oeconomy", in other words physiology. In 1795, Cuvier published no less than five memoirs with

Geoffroy, one of which is particularly notable, on a new division of mammals built on his idea, of the utmost importance, of the subordination of characters:

> Never lose sight of the fact that a type must be defined in the majority of relationships (…). It is understood that in a system as well linked as animal economics, there are organs whose conformity necessarily leads to that of most of the others, and which it must be possible to determine by reasoning and by experience (…). The life of the animal depends on the generation that gives it, and then on the regulated movement of its fluids. Generation and circulation must therefore provide the primary characters or indicators of the first order. (Saint-Hilaire & Cuvier, *Mémoire sur une nouvelle division des mammifères*, 1795, author's translation)

Thus, a carnivorous feline like the lion was defined by a set of necessarily associated characteristics that made it precisely a carnivorous feline, from the teeth to the stomach, passing through its musculature, its skeletal structure and its podal extremities. The two anatomists thus drew up a hierarchy from the general to the particular that distinguished the class (mammal), the order (carnivore), the family (felidae), the genus *(Panthera)* and the species (*leo*).

Together they created the class of "hoofed mammals" with 14 orders including the quadrumanes:

> Mammals that have thumbs separated from other fingers and that can be opposed to them by gripping. These animals are frugivorous, and essentially shaped to climb trees; they have several conformation relationships with Man; but Man is organized to stay on the ground; walk on two feet; grasp only with his hands; his posterior extremities have no separate thumbs. The narrow pelvis of quadrumanes, the compression of their thighs, the very low insertion of their flexor muscles of their leg, the articulation of their head on the neck, make bipedal walking very difficult for quadrumanes; therefore, they all have their jaws more or less prominent in the muzzle. (ibid., author's translation)

The quadrumanes included apes (simia), makis (lemur of Madagascar and indri of Southeast Asia), the lori (a lemur from Asia and continental Africa) and the tarsier (arranged with the lemurs of Southeast Asia), the two future primate groups known as simians and prosimians. Humans were absent. By deduction, Geoffroy and Cuvier considered the human as an order of mammals with hoofs. Only the adult stage was taken into account.

Cuvier acceded to the Chair of Animal Anatomy in 1802. At the age of 33 years, he was a member of the *Institut des Sciences*, Professor at the *Collège de France* and, at the Museum, responsible for the national collections of vertebrate animals.

The symbiosis between Cuvier and Geoffroy lasted only a short time: the disagreements intensified as the two ways of thinking diverged on the anatomical and physiological principles that naturally order large groups of animals.

7.2.5. *Humans have no fossil ancestor according to Cuvier*

In 1815, Cuvier resumed Buffon's history of the Earth, which introduced irreversibility in the course of the geochemical and tectonic transformations of the Earth's crust. In his *Discours sur la révolution du globe* he defended the thesis of planetary catastrophes, or cataclysms, as the natural cause of the mass extinction of species. This was the theory of catastrophism versus that of transformism. It was no longer conceivable to consider current conditions as perennial since the origin of the Earth:

> Would there not also be some glory for man to know how to cross the limits of time, and to rediscover, by means of a few observations, the history of this world, and a succession of events that preceded the birth of the human race? No doubt astronomers have progressed faster than naturalists, and the era in which the theory of the earth is found today is somewhat similar to that in which some philosophers believed the sky to be made of large stones and the moon to be as big as the Peloponnesus; but after the Anaxagoras came the Copernicus and Kepler, who paved the way for Newton; and why shouldn't natural history one day also have its Newton? (Cuvier, *Discours sur la révolution du globe*, 1815, author's translation)

This excerpt shows how topical Cuvier was:

> But what is also important to notice is that these irruptions, these repeated retreats, have not all been slow, have not all been made by degrees; on the contrary, most of the catastrophes that brought them about have been sudden; and this is especially easy to prove for the last of these catastrophes; for the one that by a double movement flooded and then dried up our present continents, or at least a large part of the soil that forms them today. (ibid., author's translation)

Quaternary paleontologists and geologists made discoveries in the great plains of Siberia subjected to global warming:

> ("The last of these disasters") has left, in the countries of the North, the corpses of large quadrupeds that the ice has seized, and which have been preserved to this day with their skin, hair and flesh. Had they not been frozen as soon as they were killed, putrefaction would have decomposed them. And on the other hand, this eternal frost did not previously occupy the places where they were seized; for they could not have lived under such a temperature. So it was the same moment that caused the animals to perish, and made the land they lived in icy cold. This event was sudden, instantaneous, without any gradation, and what is so clearly demonstrated for this last catastrophe is hardly less so for those which preceded it (...). Life on this earth has thus often been disturbed by appalling events. Unmanned living beings have been victims of these catastrophes; some inhabitants of dry land have been swallowed up by floods; others, who populated the bosom of the waters, have been dried up with the seabed suddenly raised; their very races have ended forever, leaving in the world only a few pieces of debris hardly recognizable to the naturalist. (ibid., author's translation)

Cuvier espoused Buffon's idea of the extinction of genera but he added climatic phenomena of lightning speed on large geographical scales and affecting entire populations. He posed the problem of the renewal of fauna and flora without any European academy of sciences being able to provide elements of an answer, assuming that an international college had decided to take up the issue. He did not differ from Buffon in the idea that it was new metaphysical manifestations that recomposed the diversity of morphotypes.

Nor did he conceive of primates older than humans. However, the curator of the *Hessisches Landesmuseum* in Darmstadt sent him a cast of a fossilized monkey femur with an explanatory diagram. The fossil was discovered in 1820 in the southern part of the Mainz Rhine basin near the village of Eppelsheim. The geological formations were indisputably dated to the Miocene (10 My). It was the oldest monkey fossil ever discovered. Cuvier never replied and never mentioned it (Franzen 2000). It was named *Paidopithex rhenanus* Pohlig, 1895, and corresponds to a kind of gibbon.

For the period of the appearance of humans, the paleontologist was a little too categorical; there were no fossils collected in sedimentary deposits prior to the last planetary catastrophe:

> It is certain that no human bones have yet been found among the fossils, properly speaking, or, in other words, in the regular layers of the surface of the globe (...); in the beds that conceal the ancient

> races, among the Palaeotheriums, and even among the elephants and rhinoceroses, not a single human bone has ever been discovered. (ibid., author's translation)

However, Cuvier had the opportunity to become acquainted with human remains that were different from *Homo sapiens* when he was in Stuttgart. These were the fragments of a frontal bone and a right parietal bone found in Canstadt in 1700, during excavations carried out by order of the Duke of Württemberg and preserved in the City Museum (George 1892). Fossils of Quaternary fauna had been collected in the loess and it was logical to associate them. However, rather than being interested in comparative anatomy and the hypothesis of the origin of loess prior to Diluvium, he rejected them on the pretext that they had not been collected rigorously enough to demonstrate that they were "antediluvian". However, they were Neanderthal and the assembly was coherent. Cuvier missed the first fossil of a kind of large primate very close to *H. sapiens*:

> Among the bones found in Canstadt, we see a fragment of jaw and some human works; but we know that the ground was moved without care, and that no note was taken of the various heights at which each thing was found. (Georges, *Le crâne de Candstadt*, 1892, author's translation)

In 1848, a captain of the Royal Navy, Edmund Flint, discovered an almost complete skull in a quarry in Gibraltar (the Forbes Quarry) and presented it to the Gibraltar Museum Society of which he was secretary. There was no reaction to his intervention. It was the third Neanderthal skull to go unnoticed for a long time.

Georges Cuvier may have thought of a pre-human form, but he did not rely otherwise than on stratigraphic evidence before recognizing its presence prior to Diluvium. If a first form of humanity lived in Diluvium, the races were small in number, they had drowned and their fossils were at the bottom of the seas:

> Everything leads us to believe that the human species did not exist in countries where the fossil bones were found, at the time of the revolutions that buried these bones (...) *but I do not want to conclude that Man did not exist at all before that time.* He may have lived in a few small areas, from where he repopulated the earth after these terrible events; perhaps also the places where he stood were completely damaged and his bones buried at the bottom of the present seas, except for the small number of individuals who continued his species. Be that as it may, the establishment of Man in the countries where we have said that fossils of terrestrial animals are found, that is to say, in the greater part of Europe, Asia and America, is necessarily

> posterior not only to the revolutions that buried these bones, but also to those that uncovered the layers that enveloped them, and which are the last that the globe has undergone; from which it is clear that neither from these bones themselves, nor from the more or less considerable heaps of stones or earth that cover them, can one draw any argument in favor of the antiquity of the human species in these various countries. (Cuvier, *Recherches sur les ossemens fossils de quadrupèdes*, 1812, author's translation)

He adhered to Buffon's doctrine of the unity of the human species. The variety of species remained within the limits of a general structure; it was inconceivable that natural processes could gradually transform a four-legged monkey structure into a structure made to stand upright with all the weight due to gravitation, and which ordered the bony forms as well as the joints of the posterior appendicular skeleton. He defined its great principle:

> All organized being forms a whole, a single and closed system, whose parts correspond to each other and contribute to the same definitive action by reciprocal action. None of these parts can change without the others also changing; and therefore each of them, taken separately, indicates and gives all the others. (ibid., author's translation)

7.2.6. *The division between Cuvier and Geoffroy Saint-Hilaire or the distinction between micro- and macroevolution*

The divergence between Cuvier and Geoffroy led them to become divided when the latter oversimplified the classification, reducing the history of present and extinct living beings to a primordial organization allowing for all cases by subtraction, elongation, compression or inversion. There existed only one structural plan created by the "Supreme Being", which decomposed into as many varieties:

> Nature constantly uses the same materials and is only ingenious in varying the shapes. As if indeed it were subject to first data, we see it always tends to make the same elements appear again, in the same number, in the same circumstances and with the same connections. (Geoffroy Saint-Hilaire, *Philosophie Anatomique*, 1822, author's translation)

For example, invertebrates, such as lobsters, have an exoskeleton, while vertebrates have an internal skeleton covered with muscles and organs. The distinction was made with the presence of a segmented axial skeleton that surrounded the nervous system forming the spine. Invertebrates had no solid structure that was the homolog of the internal skeleton. He made the mistake of confusing analogy with homology. Geoffroy posed it as a common feature and, first over the others, a solid frame that protected the nervous system and allowed it to move as well as the organs. For Cuvier, however, this was a confusion between analogy and homology. The segmented teguments of the articula, which formed the exoskeleton surrounding the organs, were an analogy with the segmented bone skeleton of vertebrates. The vertebrate skeleton protected the nervous system, which was the innermost of the organs, but was surrounded by muscles and organs that were not protected by armor. So the segmented and articulated spine was not the homolog of the articulated and segmented armor of the lobster that protected all the organs and muscles. It was therefore necessary to separate articulates and vertebrates and there was no common organizational plan to classify them in the same unit of composition plan.

Cuvier distinguished different structural plans according to the nervous system, arranged in parallel categories as in Aristotle's History of Animals and following a graduated rule from the least complex to the most complex. There were four branches: zoophytes (such as sea urchins), articulates (such as insects and arthropods), molluscs and vertebrates. He joined Aristotle in his laws of essences. For Cuvier, each structural plan originated in its own creation, each order was created separately; the three orders, prosimian, simian and Man, were, according to Cuvier, three independent structural plans. They belonged to the most recent of the geological periods, that which followed the last great flood of the continents, the post-Diluvium, and these three organizations were created simultaneously. For Cuvier, Buffon's bimane or Linnaeus' *H. sapiens* had no simian ancestor and therefore no primate fossils could exist.

Étienne Geoffroy Saint-Hilaire wrote and published his *Philosophie anatomique* between 1818 and 1822, with a plate comparing the skeletons of a small tailed monkey (South American alouatte), a bipedal flightless bird, the penguin and a fish (the pike) (see Plate 10 available online (Geoffroy Saint-Hilaire 1862)). This comparison paralleled the skeletons as if they had been elongated. The division between the two naturalists occurred in 1830 at the *Académie des Sciences* (Piveteau 1950).

8

The Slow Recognition of Humans' Simian Origins

8.1. Introduction

Geoffroy and Cuvier shared their questioning on the proximity of apes and humans from the point of view of psychology, a term already in use, considering the most immediately remarkable common anatomy in the possession of hands. By considering the nomenclature "quadrumanes" and "bimanes", the two anatomists avoided putting the emphasis on locomotion, and they were not mistaken. Cuvier kept his coherence centered on the nervous system, which showed an undeniable increasing complexity when passing from one branch to the other. The two anatomists examined the hand and the gripping thumb and focused their reflection on intelligence directly related to the function of manipulation:

> I mean the sense of touch. Among the organs of sensation, this sense must have without doubt the greatest influence on the whole machine, which is the most general, the most perfect, the most necessary, which alone complements the ideas acquired by the other senses. What idea would we have without it, of all that surrounds us? (...). What prodigious distance must there not be between the images that form of the surrounding bodies of a horse, for example, whose foot ends in an inert and insensitive mass, and the monkey or squirrel, whose hands are almost as divided and delicate as ours? (...) Its various degrees of perfection depend above all on the more or less pronounced division of the fingers, and their more or less delicate covering. (Geoffroy Saint-Hilaire and Cuvier, *Mémoire sur une nouvelle division des mammifères*, 1795, author's translation)

It was then that the division between the two modes of thought took all its meaning. Cuvier's belonging to the Aristotelian tradition could be understood insofar as his mind was that of a systematist, aware of the difficulty of the integral transformation of a geometry with its corresponding physics and physico-chemistry. Was this radical transformation possible during the hereditary transmission of male and female seeds? If so, then scientists were powerless to formalize it, the Newton of evolution is not yet born. Conversely, Étienne Geoffroy Saint-Hilaire understood how unreasonable it was not to question the natural links between the psychological functions and manipulative capacities of the great apes and humans. That "the Supreme Being" created two "essences" so close to each other without a common "traceability" in the past and in the face of immense geological time scales was no longer credible.

8.2. Jean-Baptiste de Lamarck or the audacity of the transformist theory of organization plans

8.2.1. *"The causes of the main physical facts" or "what is life" (1780)*

Jean Baptiste de Monet (1744–1829), knight of La Marck, was born in Picardy. He attended the Jesuit college in Amiens and, like his ancestors, he sought a military career. Forced to give up, he took courses in botany and comparative anatomy at the *Jardin des plantes* in 1771. Animal electricity had been seen in the *Gymnotus* (a freshwater fish) since 1773. Physiology, which involves the study of blood circulation, muscle contraction and pulmonary respiration, was not a new science. Lamarck acquired all this knowledge at the *Jardin des plantes*, as well as from Buffon's *Natural History of the Earth and Animals*. The decreasing complexity of living beings according to Aristotle's graduation from Man was fundamental to his teaching. The animation of organisms, whether plants or animals, was not a new theme and it was well established that the rational approach to nature proceeded from its geometric and dynamic analysis. Forces were present at the smallest scales of organization, whether they were particulate or in wave form in the case of light; universal attraction was recognized at all scales from the micro- to the macrocosm. In 1776, Lamarck set about writing a reflection on the "causes of the main physical facts" that he would not publish until 1794 (Lamarck 1794).

Parallel to this reflection, Lamarck had chosen to devote himself to a systematic study of flora for which he invented a dichotomous determination key. He wrote the *Flore française* with a foreword in which he already developed his reflections on physics and chemistry applied to botany. He submitted the work to Georges Buffon, who entrusted the foreword to Daubenton. As he was not competent, Daubenton asked the opinion of a botanical specialist, René-Just Haüy (1743–1822). As a result, the work was published in three volumes by the *Imprimerie Royale* in 1779.

However, Buffon deleted the foreword; but he supported Lamarck by introducing him to the *Académie des sciences* as Assistant in the class of Botany (the position he himself had applied for in 1736 when he was offered the direction of the *Jardin Royal*). After Buffon's death in 1789, Daubenton became the new Intendant of the *Jardin des plantes* and appointed Lamarck as "Keeper of the Herbarium of the King's Cabinet". In 1793, the young botanist was appointed Professor to the Chair "Zoology of Insects, Worms and Microscopic Animals" (Zoology of Invertebrates) and set up a program of paleontology of invertebrates.

His essay on the *Causes des principaux faits physiques* had been submitted to the *Académie des sciences* in 1780 but was not followed up, without explanation, as an echo to the censorship of his foreword. As the Academy was dissolved by the Revolution and Buffon died, the work was accepted in 1794 by the "network of resistants" with whom Cuvier was also in communication with in Normandy.

Lamarck was as multidisciplinary as his colleagues. Well-informed of works and publications, he was skilled in analysis and synthesis and was interested in the classification of the "living kingdom", from plants to humans. He took up Aristotle's degraded scale, well-enriched, and sought a unifying principle, transdisciplinary as it were, or universal, which transcended all animal and plant categories. It was this spirit of unity of the living that brought Geoffroy Saint-Hilaire and Lamarck together. This interdisciplinarity that connected them was that of the cellular metabolism and the capacity of reproduction, in other words, life. Lamarck wrote it explicitly in this text which was forbidden for publication by the Academy.

Lamarck used the neologism *biology*, created by the physician Xavier Bichat (1771–1802), to designate this new science of animate, living systems, which reproduce their organization. It was important to distinguish between physics and biophysics. The former corresponded to the laws of universal attraction which accounted for movements caused by external forces (thrust, fall, friction, whirlpools) and which did not have this property of animated and reproducible internal self-organization. A salt crystal grew, but did not reproduce, whereas the simplest microorganisms, such as Van Leeuwenhoek's animalcules, reproduced. This internal dynamism, endowed with self-replicating properties, was therefore life. No one had yet managed to explain this process.

This reflection illustrated the need to reason on the foundations of matter, following Descartes, Newton and Leibniz, with an interest in heat, but without having to refer to metaphysical considerations. The preface was in fact a criticism of the old pneumatic theories still present in the *Académie royale des sciences*, invalidated by observation and experience. They dated from Galen and maintained the dogma of vital breath, or *pneuma*. The vital breath was taken in by the nose and mouth, then to the lungs which sent it to the heart to undergo combustion in order to be infused into the blood, thus contributing to its circulation. Reaching the liver, the "vital spirit" would be transformed into a physical *pneuma*. When it reached the brain, it would be transformed into a psychic *pneuma*, which would convey the sensations perceived by the five senses.

Aware of the novelty of his synthesis, based on observations in botany, he specified that, for protection of his anteriority, he took care to have his publication signed by a member of the Academy. Indeed, his thought was that of a physicist who had the audacity to understand that physics and chemistry were to be integrated in the study of internal movements within and between cells:

> Causes of the main physical facts and particularly those of Combustion, of the rise of water in the state of vapors; of the Heat produced by the friction of solid bodies with each other; of the Heat which makes itself sensitive in sudden decompositions (…); finally of the maintenance of the life of organic beings, of their growth, of their state of vigor, of their decay and death (…). Delivered now from all anxiety, by the benevolence of the French Revolution, I do not hesitate to publish my work today (…). Life is constantly communicating by the multiplication of individuals who are equipped with it, who pass it on to others, and who continually succeed one another (…). Long meditations on the subjects I have dealt with, and many observations that are related to them, and that belong to me, have given birth to this work that can be called a kind of physico-chemical logic (…). (Lamarck, *Recherches sur les causes des principaux faits physiques* (…) *l'entretien de la vie des êtres organiques*, 1794, author's translation)

The full title already gave the measure of the necessity of a new science on the physico-chemical studies, which were organic animation through heat production, the fight against death and disorganization. With a generation's gap, Sadi Carnot had understood this difference between the machine built by a human's hand, which did not reproduce itself, and life; his essay on the thermodynamic forces appeared two years before the zoologist's death.

8.2.2. *Transformism or the first formulation of evolution*

Lamarck thus presented the present era as the culmination of the development of extinct organisms according to a graduation from the simplest to the most compound. This evolution was formulated in a small book of 146 pages entitled *Recherches sur l'organisation des corps vivants et particulièrement sur son origine, sur la cause de ses developpemens et des progrès de sa composition*, signed

J.B. Lamarck, member of the *Institut national de France*, Professor administrator of the *Muséum d'Histoire Naturelle, Sociétés d'Histoire Naturelle*, etc., in Paris, at the author's home at the *Muséum d'Histoire Naturelle*.

The zoologist explained that the work was not a project; he had the idea following his introductory speech delivered on the revolutionary date 27 Floréal year X (May 17 1802) for his classes of the same year. He had written his lectures on loose sheets that he proposed to distribute to interested students and then he chose to develop his reflection:

> I am well aware that the novelty of the considerations set out in this paper, and above all their extreme dissimilarity from what is customary in this respect, required more extensive development so that the basis of the considerations in question could be better established and more easily seen. (Lamarck, *Recherches sur l'organisation des corps vivants*, 1802, author's translation)

Taking the scale of animals in the order of Aristotle or the "Degradation of organization from one end of the animal chain to the other", he wrote:

> This degradation imperceptibly leads to the inconceivable end-point of animalization, that is to say, to one where animals are most simply organized, in a word, one inhabited by those suspected to be barely endowed with animality, who in fact are probably the first drafts of it, and undoubtedly it is this 'end' by which nature in fact began, if it is true, with the help of much time and favorable circumstances, and which then succeeded in forming all the others. (ibid., author's translation)

Since Lamarck was convinced of the past reality of an evolution, under the influence of Georges Buffon in spite of everything, he postulated a mechanism. For what reasons would the organization of a plant or an animal change if its environment does not change? However, living in its environment consisted of developing interactions through different functions. If therefore the organs and skeletal structures had changed, it could only be under the influence of functions, hence the famous formula "the function creates the organ". He set out six points, the fourth and fifth of which constituted his founding idea of the mechanisms of change and the filiation between different species: "4 – Organization in every living body was obtained little by little by the progress of the influence of the movement of fluids (…) 5 – each organization and each form acquired by this order of things and by the circumstances that contributed to it were preserved and successively transmitted by generation, until new modifications of these organizations and forms were acquired by the same way and by new circumstances" (ibid., author's translation).

He set out his main principles:

> To change each internal system of organization requires a combination of circumstances more influential and of much longer duration than it does to alter and modify the external organs. Nevertheless, I observe that when circumstances require it, nature passes from one system to another, without making a jump, provided they are neighbors. It is in fact by this faculty that it has succeeded in forming them all successively, proceeding from the simplest to the most compound. In animals with udders, which are the most perfect, we see a sexual generation constituting viviparous animals: an immediate succession of vital movement in the embryo, which it has just received at fertilization; a nutrition of the fetus during its first development at the expense of the mother's substance, with which it does not cease to communicate until birth. (ibid.)

Lamarck introduced intrauterine life and the relationship between mother and fetus and he did not exclude a possible influence on the formation. On the other hand, he did not comment on the physiological, or physico-chemical, mechanism that would transmit so many changes in coordination, in the oocytes or spermatogonia of the developing fetus or after birth.

He set out a series of postulates that followed on from one another:

> It is not the shape of the body or of the parts that gives rise to the habits and the way of life of the animals, on the contrary, the habits, the way of life and all other influential circumstances have with time constituted the shape of the body and the parts of animals. The main circumstances arise from the influence of the climate, of all the surrounding environments. Constant lack of exercise of an organ gradually impoverishes it and eventually destroys it.
>
> As a result of these diverse influences, the faculties are extended and strengthened by use and diversify, and imperceptibly this conformation participates in the consequences of all these influences, it is preserved and propagated by generation.
>
> I give the name of *species* to any collection of individuals which, over a long period of time, are so similar in all their parts that these individuals present only small accidental differences that reproduction causes to disappear (…).

> After a long time, the totality of individuals of such species changes along with the circumstances that act on them, they take new forms as a result of a long habit and then they constitute a new species. (ibid.)

In the course of his writing, he did not fail to criticize those who claimed to know better than anyone else that this was a matter of divine will rather than natural mechanisms.

His ideas were well formulated from the title of the first part: *Des progrès de la composition de l'organisation des corps vivans, à mesure que les circonstances les favorisent.*

8.2.3. *Lamarck and the first theory of the common origins of the orangutan, the chimpanzee and* Homo sapiens

Lamarck thus understood the need to find a fossil ancestor common to *Homo sapiens*, the chimpanzees of Angola and the orangutan of Java (*Pongo*). He extrapolated his idea to Daubenton's work on the skull, with his conclusions on the differences in skeletons. He integrated Geoffroy's and Cuvier's reflections on sensory intelligence developed with the touch and sensitivity of the fingers, including their joints adapted to gripping:

> As for the human closest to the quadrumanes, he is strongly distinguished by a great superiority of intelligence and also considerably by several organizational features which characterize him: 1) the occipital hole completely in the base of the skull, which makes his head rest on the end of the spinal column and makes him see a much greater number of objects at the same time; 2) the remarkable mobility of his fingers; 3) the state of his organization, the man is made to stand and walk upright. Quadrumanes do not have the occipital hole located quite in the base of the skull, but they certainly have much less raised it posteriorly than all other mammals. Therefore they are more often standing upright, although this is still very awkward for them. The Angolan Kimpézé (chimpanzee) usually walks upright, leaning on a tree branch as a stick. Wouldn't one think that this state of organization has been acquired little by little over a long period of time, with the help of circumstances that have been favorable to it? The way of life and all other influential circumstances have, over time, formed the shape of the animals' bodies and parts. The main circumstances arise from the influence of the climate, from all the surrounding environments. (ibid., author's translation)

8.2.4. *The anatomical origins of* Homo sapiens*, a break with his own statements*

The naturalist broke with his own logic that defined the order of the criteria of distinction from the inside of the organism, which was first, to the outside, which was second. He had not studied a primate skull himself, so he was unaware of its osteological organization and, above all, like the rare primatologist anatomists, he was unaware of its internal morphological organization. Locomotion was not the primary criterion of the very few anatomists who had had the opportunity to compare the chimpanzee, the orangutan and humans. The organization that brought them together was not anatomical, it was physiological, with the nervous system and the fingers, and behavioral. Human intelligence was related to the verticality of the nervous system and its organization and not to bipedalism, which is common in primates. However, nothing was described concerning the cranial bone structure, which is closely related to verticality and balance control, involving the position of the cerebellum and especially that of the brain stem which begins with the straightening at the center of the cranial base.

Daubenton, Geoffroy and Cuvier did not focus the significance of the differences between quadrumanes and bimanes on the locomotion of the lower limbs; this difference was never causal. It was systemic and Cuvier understood it very well. The principle of correlations was of very great relevance: it was never said that the occipital hole of humans was lower and further forward than an ape's "because Man walks", it was written that it was the whole that was made to walk standing upright. Humans could not move other than on their two legs because of this conformation.

Lamarck's causal sequence for the origins of human anatomy was very clear. The process postulated was: 1) under the effect of climate, postnatal locomotor functions modified the internal organization of the skeleton and organs; 2) macroscopic changes were memorized in sex cells at molecular or even atomic levels; 3) these new molecules were transmitted from generation to generation between males and females at population level, because conditions remained favorable. If they were no longer favorable, individuals died and no longer reproduced. The hypothetical mechanism of a change, and of its transmission, was thus well explained; it was climatic conditions that, at the same time, had transformed anatomy and would have allowed transmission through reproductive pathways very gradually with each memorization in the sexual cells. If the initially favorable climatic conditions were maintained, or evolved in the same trend, the process would be maintained and transmitted over millions of years.

Buffon's terrestrial cooling found its application. Permanent bipedalism would be the substitute for arboricolism following the rarefaction of forest cover, under the effect of the lowering of global temperature. This doctrine was clearly a development of Buffon's concepts, "revolutionized" by Lamarck, who felt free, with the French Revolution, to transgress the prohibitions of theologians and to provide a physico-chemical reality, where Buffon had been forced to invent the "internal mold" and "passive organic molecules".

When Lamarck imagined the straightening of the axial skeleton from a prosimian one (the antero-posterior polarity remaining horizontal) to a human one (the axial polarity being mostly vertical), he limited himself to the spine and the occipital hole, ignoring the internal organization of the cranial base. It was not described in primates, neither by Daubenton, from whom he drew his inspiration, nor by any anatomist who had compared the skulls of apes

The invertebrates specialist did not propose mechanisms capable of accounting for the reorganization of bone structures ordered from the head to the last vertebra, nor that of the organs associated with the change that requires consideration of gravitation, such as the heart and the cerebellum, for example, and, above all, he did not give an explanation for the transmission of adult geometric and organic changes to the cells of reproduction.

In other words, Lamarck would not have grasped that, for Cuvier, the differences between the great ape and the human being were at the level of order and not that of the small variety of species with the color of hair or the number of vertebrae of the caudal appendage.

Lamarck had to develop a system of thought capable of explaining the complete reorganization of the adult organism. Beyond that, for a naturalist who excluded metaphysics from the field of science, it was imperative that the infinitely small cells of male and female reproduction be compatible with the laws of attraction.

Cuvier can be seen as an irrational naturalist; nevertheless, Lamarck's vision was no less irrational in believing that muscular constraints on the scale of the adult body could have not only transformed the entire organization of physiology, both outside and inside the axial skeleton, but that they could also have been inscribed in the seeds of males and females in exactly the same patterns. Charles Darwin was the great continuator of this mode of reasoning.

8.2.5. *Which system to classify humans: the separation of organizational plans or the variety of a single plan?*

Lamarck expanded his theory in 1809 by comparing invertebrates, which he mastered best, with vertebrates. Among the invertebrates, which ranged from corals to lobster via sponges, oysters and earthworms, the animals that brought the vertebrates to mind at first glance were those organisms that had the same asymmetrical geometry in the cephalocaudal sense.

All had a head aligned on the axis of the body, whether an earthworm, an ant or a shrimp, and this organization was segmented in the antero-posterior direction. The simplest organization was the repetition of the same pattern, such as the ring of the earthworm, an invertebrate, or the vertebrae of the vertebrate. This repetition was polarized in the antero-posterior or cranio-caudal direction, the front being defined by the direction of movement of this axis. Animation was under the control of the nervous system, which centralized information and allowed the organism to move, with or without locomotor limbs, like the earthworm or snake moving horizontally with its spine.

Georges Cuvier was categorically against Lamarck's vision, which saw only the surface of organizations, forgetting the structure of the internal organization, especially the skeleton. Although he was not aware of the organization of the internal base of the skull, Cuvier did not conceive of its prosimian transformation into human organization under the effect of bipedal locomotion, which was more frequent than in quadrumania; the structures were coordinated and they could not be transformed simultaneously. External constraints could not transform the internal organization set up very early on.

Étienne Geoffroy Saint-Hilaire, on the contrary, adhered to Lamarck's speculation, since for him the idea of progressive variation was permitted by nature within the limits of a single plan of organization created by the Supreme Being. He broke with two millennia of Aristotelian essentialism: relations between types were conceived through changes in the conditions of existence.

Following in the footsteps of his colleague Jean-Baptiste Lamarck, Geoffroy Saint-Hilaire placed comparative anatomy at the center of the biological sciences. For Cuvier, classification was an end in itself, whereas for Geoffroy Saint-Hilaire it was a means to understand the organization of transformist mechanisms.

It was a threshold of maturity unique since the philosophers of Antiquity; Geoffroy Saint-Hilaire reversed the roles. Fixist theologians and creationists were kept out of the field of natural sciences. Then, the theory of evolution gave anatomy and comparative physiology the central place in continuing the progression of knowledge.

Lamarck introduced the necessity of transformism to make the interpretation of natural history of fossil and present-day species coherent. However, for the origins of monkeys, great apes and humans, his vision revealed a profound misunderstanding of the internal structures of the skull that condition the position of the occipital hole. The constraints came from inside the axial skeleton, although there were no muscles. The invertebrates specialist did not want to make an exception for humans and therefore chose the easy way. His vision was accessible to the intuition of common sense, but it was nevertheless an illusion that mistook effects for general causes. Bipedalism had never been the criterion for classification and it had no causal relationship with the straightened organization of the axial skeleton of humans.

Georges Cuvier died on May 13 1832, leaving his classes at the *Collège de France* unfinished. He was replaced by the physician and paleontologist Henri-Marie Ducrotay de Blainville (1777–1850). Trained by Cuvier, de Blainville was immediately appointed holder of this chair but it did not fall within his competence; so the day after the master's death in 1832, he took over the Chair of Comparative Anatomy. It was his publisher who created the neologism *paleontology* in 1834.

De Blainville was not a transformist but he disagreed with the conception of successive re-creations (catastrophism of Cuvier) without graduation. As such, he was in favor of a single creation with the successive appearance of organizational plans, the great majority of which would end up in a fossil state. In fact, it seemed conceivable to him to collect apes and fossilized humans from before Diluvium. In 1839, Johann Wagner (1797–1861), Professor of Zoology and Paleontology at the University of Munich, discovered in southern Greece, near Athens, fossils of an ape similar to the macaque in the Miocene paleontological site of Pikermi (7–5 Ma). He named it *Mesopithecus pentelicus* (Wagner 1839).

The new doctrine of Étienne Geoffroy Saint-Hilaire also foresaw the successive appearance of pre-human morphotypes that moved away from the great ape and closer to humans, whose anatomy came to be defined by *sapiens*. The evolution was not finalized by the anatomy of *sapiens* since it was constrained by the environment. Paleontologists had to be careful not to speak of human characters, since humans did not exist; it was humans who preserved characters of these morphotypes still unknown. However, the locomotor and graduation straightening that Lamarck postulated were not credible for anthropologists. It was through another cause that embryologists interested in the transformism theory attached themselves to transformism.

Comparative embryology searched for the causes of the process by tracing the organogenesis back to fertilized eggs. These were the schools of thought that associated ontogenesis and phylogenesis, involving Caspar Wolff (1734–1794), Johann Meckel (1781–1833), Karl von Baer (1792–1876) and Ernest Haeckel (1834–1919).

9

Embryology, Fixist Anthropology and the Neanderthal Man

9.1. Introduction

Scientific and conceptual research was progressing in paleontology, physiology and embryology with the German anatomists Johann Meckel and Karl von Baer. New living species and fossil skeletal fragments of unknown genera or families complicated classification. The horizontal linear representation had given way to representation as a tree with branches diverging from the branches that gave rise to them via successive dichotomies. However, the general trend remained in the complexification of the nervous system, which began with embryogenesis and whose contemporary optimum in verticalized *Homo sapiens*, with its capacities for abstraction, its works of art and its funerary rituals, was not contested by any paleontologist. No anthropologist, naturalist, paleontologist or specialist in comparative anatomy was surprised by this.

The generation of species thus became the challenge to support or invalidate the idea of evolution. Ovism and animalculism postulated the complete formation of the organism in the spermatozoa or eggs of females. These two reasonably unsustainable postulates, as Georges Buffon put it, were abandoned in favor of the theory of epigenesis developed by the German biologist Caspar Wolff. Epigenesis was quite simply ontogenesis, the development of the organism from the single cell stage followed by growth by cell division with the differentiation of cells, increasingly specialized in their future organic functions, which were complementary and vital to the overall metabolism of the organism.

9.1.1. *The theory of epigenesis (Wolff 1759)*

Caspar Wolff (1734–1794) was born in Berlin. He learned surgery, anatomy, microscopy, chemistry and botany at the Medical-Surgical College of the Royal Academy of Sciences in Berlin. The college was directed by Pierre de Maupertuis, the mathematician who had appreciated the first early essays of Georges Buffon. Wolff continued his university studies at the Faculty of Medicine of Halle University with a thesis defended in 1759, entitled *Theoria Generationis*. It was completely innovative and aroused a most justified interest. Wolff took up Aristotle and William Harvey's theory on epigenesis.

Using a microscope, Wolff was able to observe the embryonic development of plants and chickens by describing them with very precise drawings. The stages of development of a chicken were described from the embryo visible on the egg yolk to the formed chick. The conclusions challenged the dogma of pre-formation supported by the well-known European Bernese physician and naturalist Albrecht von Haller (1708–1777).

Haller read the thesis and rejected it, following converging observations in botany and zoology that invalidated it. In 1767, the Imperial Academy of Sciences of St. Petersburg nevertheless entrusted its Chair of Anatomy to Caspar Wolff. Wolff continued his research and described the formation of layers that he called "germinal layers". He published them in 1768 and 1769 (Wolff 1768–1769), demonstrating, if still necessary, that it was possible to follow the stages of a differentiated and growing organism after fertilization. The idea of epigenesis was thus validated; the organism was not contained in miniature in the eggs of females or the spermatozoa of males. It was an organism created with fertilization.

9.1.2. *Karl von Baer discovers the formation of the ovum (1827)*

Karl von Baer, born in 1792 in Estonia, was the originator of comparative embryology. In 1817, he joined the University of Königsberg (Northern Germany on the Baltic Sea) and became Professor of Zoology in 1821, then of anatomy in 1826. In 1827, he studied the ovaries of dogs under a microscope and discovered the formation of the ovum. De Graaf had identified follicles in rabbits in 1672, not the laying of the egg to be fertilized, while Wolff had described the stages of embryo development. What was missing was the succession of stages from follicles to the embryo that had been formed. The formation of the oocyte had never been observed and had never been followed up until the implantation of the fertilized egg in the wall of the uterus.

Van Baer's observations were published in Latin in 1827 by the Imperial Academy of Sciences of St. Petersburg under the title *Ovi Mammalium and Hominis genesi* ("On the Genesis of the Ovum of Mammals and Man"). The formation in the ovaries was detailed, the plates engraved and colored by hand, illustrating all the stages up to the well-advanced stage of the embryo at the stage of cervical flexure; it was a late and provisional bend between the future brain and the future spinal cord (Baer 1827).

This study therefore represented a historical step forward because, since De Graaf, the formation of reproductive cells, and therefore their origin, remained imprecise and the link with the embryo implanted in the uterus was hypothetical.

Von Baer therefore completed the stages, starting with the ovaries, observing the formation of follicles from which the oocytes were expelled into the Eustachian tubes. This was followed by the fertilization of the oocyte by the sperm, which became a zygote. Cell division began in the fallopian tubes in the morula and blastula stages before implantation in the uterus in the gastrula stage (see the color drawing available online (von Baer 1867)).

The evocation of "Man" in the title of the article was confusing because the description of human embryogenesis was not within its scope. It was then being studied, in France, Paris to be precise, by Alfred Velpeau. He embarked on a very long-term study which required the establishment of a collection of 200 human fetuses less than three months old. It was not published before 1833.

Embryology made it possible to get closer to the unity of composition plan so dear to Étienne Geoffroy Saint-Hilaire; but what would it reinforce: the insensitive gradualism within the limits of the same plan or the structural discontinuities?

9.1.3. *Johann Meckel (1781–1833), the revolution of the 11 laws of embryogenesis*

Johann Meckel (1781–1833) was born in Halle-sur-Saale (East Central Germany). His childhood was steeped in a world of medical curiosities collected by two generations of obstetricians, pathologists and anatomists. The family home was a first-rank museum in Europe. His father intended him to pursue this vocation, initiating him very early on into the preparation of anatomical specimens. The young Meckel thus blossomed among skeletons and alcohol jars containing organs and a quantity of "human monsters". He followed the lessons of the physician Johann Blumenbach (1752–1840), specialist in natural history and anthropology (Blumenbach 1795, 1798). Blumenbach was responsible for the word "race". He defended the theory of monogenism, which posited a single strain at the origin of ethnic diversity.

In a European intellectual context where biblical doctrine was still very influential, the only conceivable theory for understanding biological diversity was the influence of climate and living conditions. Buffon was bold enough to conceptualize it in 1749 and Lamarck generalized it to the whole order of primates in 1802, overlooking the problem of structural limits between the morphological organization plans, *H. sapiens* not being a variety of great ape. Meckel obtained his Doctor of Medicine degree at the University of Halle in the same year, 1802. His thesis dealt with heart malformations in specimens from the family collection. With his doctorate in hand, he went to Paris in 1803 to collaborate with Étienne Geoffroy Saint-Hilaire and Georges Cuvier. In 1806, the University of Halle was occupied by Napoleon's armies. The Emperor therefore appointed Meckel associate professor and asked to visit the family collection. He was opposed to this. In retaliation, the university was closed and the family home became the headquarters of the French army staff until its departure in 1807.

When the university was liberated, Meckel was immediately appointed to the Chair of Anatomy. He continued the study of embryonic development, in particular the nervous systems of different mammals, and devoted himself to the translation into German of Cuvier's *Leçons d'anatomie comparée*. He completed them with his own observations on malformations and published them in 1810 in five volumes. He noted the persistence of embryonic structures in the human fetus, among them "Meckel's cartilage". This cartilage is the embryonic mandible to which we will return in Volume 2, demonstrating that it was the key to the relationship between flexion of the base and straightening of the lower face (the mandibular symphysis).

Meckel described 11 laws governing the form of organs, the eighth of which stipulated that *no organ has exactly the same qualities at all times of its existence.* His theory was built on the comparison of embryos and fetuses, which allowed him to put forward a theory of major importance.

The previous state of development of a less elaborate animal organization was integrated into the embryonic development of the derived organization; the new organization was not an addition. The equation was not an addition.

Along with Johann Meckel, Étienne Geoffroy Saint-Hilaire understood the importance of malformations in fetal development to infer the natural causes responsible for morphogenesis. The comparative embryology of vertebrates and invertebrates would perhaps give reason to his *theory* of the unity of the *plan* in organic *composition*. Discussions on the homology of organs, and of bones, were therefore necessary to give transformism its empirical framework.

In 1819, Isidore Geoffroy Saint-Hilaire (1805–1861), the son of Étienne Geoffroy, studied several cranial deformations of human fetuses as well as brainless "monsters" which he published in 1822. He made it a discipline that he called "teratology" and he classified the malformations into categories, but they did not go beyond the morphological limits of humans.

Embryonic and fetal malformations remained developmental abnormalities involving the excess or lack of organ differentiation; they did not demonstrate a simian phase in a defect of organogenesis.

Observations did not go back far enough in development to be able to note a hypothetical state of ape organization and support his father's thesis, which implied an undifferentiated period common to all current primate species, including humans. Moreover, the fetal stages of apes needed to establish comparisons between each stage were non-existent in the collections.

Nevertheless, teratology was of interest in that it indicated possible mechanisms of change, having understood that it was only during embryonic formation that structural and organogenic modifications were possible. In fact, the "monsters" indicated that modifications were possible during morphogenesis; they were alterations and they could perhaps provide information about what was altered and therefore susceptible to modification. This was already a considerable contribution of knowledge supporting the theory of transformism at the scale of organizational plans, a theory rejected by Georges Cuvier.

The work of von Baer moved in the direction of the transformation of structures when he enunciated in 1828 the *principle of ontogenetic precedence of homologies, or the law of differentiation of the rough outline*. Homology was the keyword that divided Georges Cuvier and Étienne Geoffroy Saint-Hilaire. The articulation able to give reason to the two protagonists found its foundation in the understanding of embryonic morphogenesis. Cuvier was right not to follow Lamarck's overly simplistic reasoning and to insist on the discontinuities between structures; they required explanations much more complex than "function creates the organ" – heredity was left out. However, Geoffroy Saint-Hilaire was right to insist on the possibility of a modification of homologous traits by natural means on condition that it started from the beginning: the construction of the organism from embryogenesis.

Obviously, the question concerning the hereditary memory of these morphogenesis remained unanswered. It was somewhere in the oocytes and spermatozoa, on scales that escaped the microscope and in a form that defied imagination, if not understanding. Georges Cuvier could not be wrong about the substance; the corpus of knowledge was powerless to answer the most fundamental of questions concerning the origin of the vertical anatomy of humans.

The debate between Cuvier and Geoffroy Saint-Hilaire went beyond the French borders and fascinated Europe. The German poet Johann Wolfgang von Goethe had a passion for natural history and was familiar with the "Meckel" collection.

9.2. The origins of the vertical anatomy of humans: between poetic metaphysics, transcendental finality and climatic influences

Since the reaction of Jacobus Sylvius against André Vésale, anatomists have understood that apes and humans correspond to different structural and organogenic plans. Among the points that separated them, Vesalius had retained the intermaxillary bone of the palate, which is non-existent in adult humans. This difference brought an additional argument to the theory of a distinct metaphysical essence in the quadrumanes and the bimanes.

Goethe was rather sensitive to Leibniz's metaphysics, which identified the macrocosm as the living body of a cosmic organism, and he appreciated his analogy of metamorphosis, although the idea of evolution was absent in Leibniz's work.

The cosmos was not yet seen as a cosmogenesis; it was not an organism under construction with billions of multiform creations in the Earth's oceanic placenta and in the amniotic liquids that keep its vital aquatic necessity. The future of the cosmos and of *H. sapiens* were not yet themes for reflection, they were still unthought-of things that would end up being discussed at the *Muséum national d'Histoire*, but not before the middle of the 20th century.

Goethe had sensed the simian origin of humans since 1784, supported by the human remains of Canstadt discovered in 1700, while Cuvier had preferred to cast a modest veil over these disturbing old bones. The poet therefore had the idea of verifying whether intermaxillary bone existed in the human fetus or, more precisely, whether the sutures that individualized it were visible at an early stage of its development. He was able to find this out thanks to the "Meckel" collection. The observations were published in 1784 (Goethe 1784) and he took them up again in 1831 (Goethe 1831) after learning of the polemic between Cuvier and Geoffroy Saint-Hilaire. Geoffroy Saint-Hilaire therefore had Goethe's full sympathy.

The criterion for classification at the genus, family or order level depended on an understanding of the organization plan that unified or separated the species being studied.

The debates therefore did not come about as the result of the radical disagreement between the two professors of the Museum. This disagreement concerned the finality of animal forms in a universe which had no history.

For Cuvier, the finality of an animal form fell within the limits of an organizational plan common to several species that could be recognized by homologies. The organs and the skeleton complemented each other and reciprocally returned their physiological and supporting functions to each other; they finalized each other. The reciprocity of functions, or their finality, was part of the dynamic and viable coherence of the whole living system and had no reason to change.

The human morphology organized around verticality could not be interpreted as the revised and modified organogenesis of a quadrumane great ape. For Georges Cuvier this was inconceivable and unjustifiable. The reciprocal functions or purposes of the organs and supporting tissues could not be modified in their uniqueness, with the new purpose of walking upright. For this man of science, the nervous system should never be lost sight of; it was first in every respect, from its morphogenesis to its cognitive properties. Structures and organs were differentiated to allow the nervous system to move and function.

The state of the art was therefore as follows: the Earth had a history in a cosmos that had none. The dilemma was therefore considerable, since, according to the Bible, unlike other species, humans had a destiny not for themselves, but in relation to the first cause of the universe; their destiny was linked to this first cause. The vertical structure of humans was thus understood as a finality of cosmic origin and not as a plastic adaptation of the adult body to the slow disappearance of trees, as Lamarck postulated. Verticality was not a locomotor adaptive consequence, nor was it a feature of the skeleton; it was primarily a feature of the central nervous system. For the vast majority of naturalists, the verticality of the nervous system and the cognitive capacities of humans, such as conscious reflection and access to mathematics and the laws of nature, were part of this purpose of an unchanging cosmos.

It was around the notion of finality that conflicts crystallized, more because of the dogmatic rigidity of theologians, who intended to govern the life of each human being, than because of a plastic and poetic metaphysics as Goethe knew how to transcribe it, in the same vein as Leonardo da Vinci.

In the poet's eyes, this metaphysics appeared as an ode to life and not as a depressive melancholy in the face of the corruption of the world. The analogy was the shedding of the chrysalis to reveal a butterfly, from the dynamic form to its completed essence, which allowed the soul to reach its metaphysical finality. Emotion and feeling remained the ultimate force that lifted it toward its primary cause:

> Inanimate objects, do you then have a soul that clings to our soul and the strength to love? (Alphonse de Lamartine, *Milly ou la terre natale*, 1830, author's translation)

9.3. Great confusion between Linnaean nesting classification and the emergence of organizational plans

Evolutionary paleontology and embryology complemented each other in their different branches and it was conceivable that, in the end, the evolution of organizational plans would be that of embryogenesis, from which new subspecies would diverge, with variations in fetal development and then post-natal development. They would become species, and as the temporal divergence progressed, the traits would "settle" into the category of sub-family, family, infra-order and then order. However, this reasoning failed when it was no longer able to account for the emergence of a new morphogenesis along the cranio-caudal axis of the embryo. This could no longer be considered as a subspecies, or a variety, of the recomposed plan. This was precisely the phylogenesis of the first *H. sapiens* primates. Phylogenesis was a succession of embryogenesis and restructuring correlated to the strengthening of the dorsal cord. This question of embryonic organization was therefore of paramount importance in human paleontology and for its development in the 21st century.

The schism between Étienne Geoffroy Saint-Hilaire and Georges Cuvier was thus the prelude to the great confusion between the levels of organization that it made sense to distinguish from the first fossils resembling *H. sapiens*. Both were right. Geoffroy Saint-Hilaire got it right, Transformism is a past reality at the origin of the long human lineage from a strain of primates whose closest current species are the Lemuriforms. However, Cuvier was also right, human embryogenesis does not pass through a stage common to the current lemur, then to the current macaque, then to the current chimpanzee. Thresholds of complexity were reached and crossed. No one was able to explain how, and this is still true in the 21st century.

9.3.1. *Étienne Serres (1786–1868) and the "transcendental" anatomy of the embryo (1832)*

Human embryology gradually developed at the *Muséum national d'Histoire naturelle* with Étienne Serres (1786–1868), who studied and dissected "monsters" on behalf of Étienne and Isidore Geoffroy Saint-Hilaire. Serres had practiced medicine since 1810 and pursued a career as a researcher in pathological anatomy at the Paris Faculty of Medicine and the *École anatomique des hôpitaux*. The study of teratogenesis fed reflection on the mechanisms of embryogenesis. Although they could not be studied *in vivo*, they were nonetheless induced from different fetal malformations as a hypothesis. The aim was to understand morphogenesis in a move from the simple to the most differentiated.

In 1832, Étienne Serres published a description of several cases of conjoined sisters or brothers with different degrees of twin fusion. He dedicated his book to "Étienne Geoffroy Saint-Hilaire, President of the *Académie des sciences* and professor-administrator at the Museum" (Serres 1832) and proposed a theory of organogenesis (embryogenesis) integrating the concept of *transcendent anatomy* of Henri de Blainville: "Transcendent anatomy, that which, from the observation and comparison of concrete anatomical dispositions, rises to the abstract conception of the laws of organization considered in its various degrees":

> The research on organogeny that I have been doing for several years has allowed me to establish the rules that nature follows in the successive formation of organs (…). *The embryo of all vertebrates is formed and develops according to these constant laws.* Any embryo, or any organ that does not reach its proper type, does so according to a derivation of these rules. Any organ or embryo that exceeds this type is also subject to them (…). The more we study the embryos of animals and man, the more we find that the formation of organs is gradual and successive (…). I gave the name of *transcendental anatomy*, in the work on the laws of osteogenesis crowned by the *Académie des sciences* in 1819 (…) so exactly described and defined by M. de Blainville who wrote: *transcendent anatomy seeks to account for the increasing or decreasing composition of animals, its main purpose is really physiology, or the explanation of the phenomena of life, by the general application of the laws of nature* (…) We have seen the tendency of the bony, vertebral and cranial system, to reproduce in their deformation, the simple organization of the sacrum. This regression from the most compound organs to the simplest organs is found in many other cases. (Serres, *Recherches d'anatomie transcendante et pathologique*, 1832)

He described spina bifida, or the lack of closure of the spinal cord and vertebrae, which concluded the cessation of a dynamic process whose trajectory was the closure of a gutter in tube. These dynamics were at the heart of the straightening of the dorsal chordate specific to the primate lineage.

Étienne Serres held the Chair of Anthropology from 1839 to 1855, then the Chair of Comparative Anatomy, which was left vacant following the death of Louis-Georges Duvernoy. Duvernoy had some time ago succeeded the paleontologist Henry de Blainville, who died in 1850, while continuing to teach at the *Collège de France*. De Blainville, Cuvier's successor, had hired an embryologist at the Museum, Victor Coste, in order to develop comparative embryogenesis. The Chair of Anthropology was entrusted to a strong personality whom we will meet again in discussion with Charles Darwin. It was Armand de Quatrefages, the inventor of the Cro-Magnon race.

9.3.2. *Alfred Velpeau (1795–1867) and the cranio-caudal gradient of embryogenesis (1832)*

Alfred Velpeau was a physician-surgeon of modest origin. He published a titanic work, unique in the world, the fruit of eight years of collecting observations on the first two months of *human embryogenesis*. Velpeau was born in a village in the Tours region (France). His father was a village blacksmith who provided veterinary care; the young teenager discovered medicine in one of his books and decided to make it his profession. By dint of perseverance, he learned to read and speak French. He became a doctor of the Faculty of Paris in 1823, then an experienced surgeon and obstetrician, which earned him the Chair of Clinical Surgery from 1833 to 1867, and he was elected member of the *Académie de médecine* in 1832 and of the *Académie des sciences* in 1843. Velpeau delivered babies and had as many "human eggs" as possible (placenta, umbilical cord and embryo), provided by doctors and midwives. He had been interested in the generation of the embryo since 1821 and brought his observations to the attention of academics as early as 1824. He published an important synthesis of these observations in 1833. It was a remarkable study on "Human Embryology or Ovology, containing the Descriptive and Iconographic History of the Human Egg", illustrated with 15 plates.

Velpeau gathered his observations at the same time as von Baer. The French anatomist was very critical of anatomists who extrapolated on human embryogenesis with so few observations and their lack of method and rigor. Human embryology was not a constituted discipline because it required time, contacts and exceptional conditions to constitute a series of embryos and fetuses expelled with an optimal guarantee of reliability:

> To demonstrate a fact in human embryology, conditions are indeed so rare that they are nowhere to be found. (Velpeau, *Embryologie ou ovologie humaine*, 1833, author's translation)

The surgeon discussed the obstacles of such an undertaking: from the observation of the most microscopic stages to the delicate dissections of "the egg" (the placenta and its contents), the uniform color of the tissues did not facilitate a distinction (e.g. the vessels of the umbilical cord). He had to collect 200 cases, recognizing his own errors, which put him in a position to make severe criticisms of those works that claimed to describe embryogenesis. He noted that spontaneous expulsions were in the majority in the third month and that they were the result of malformations which occurred at this stage. He cited about ten unreliable authors whose descriptions could not serve as examples. Consequently, he chose to describe the development of the first week of pregnancy and to stop before the 12th week, or the end of the third month. He progressively drew the best preserved specimens, specifying whether the scale was natural or enlarged.

The drawings were of varying quality, some reproductions of authors (Hale or Sœmmering), others the originals of his collection. The different stages of the embryo were dispersed between the plates, and only the general profile was drawn, or even contoured, with the linear or coiled shape of the neural tube. The youngest embryo mentioned was 12 days old; he did not give an illustration of it. The next stage at three weeks was drawn and was already instructive; his interpretation announced the recognition of a cephalo-caudal gradient in the differentiation of the embryo (according to the head-last vertebra direction):

> Ovum about three weeks old. The cerebrospinal stem forms the base of this fetus. Its thorny crest was already visible, the nose, eyes, mouth, neck, arms, legs, also have their plane forward. It is also easy to sense that all the viscera will develop in this direction. (ibid., author's translation)

This gradient was topical in 1983 in the embryonic genetics of fruit flies and mice. The discovery of the segmentation genes regulating this gradient in Drosophila earned Christiane Nüsslein-Vothard, Eric Wieschaus and Edward Lewis the Nobel Prize in Physiology or Medicine in 1995 "for their work on the genetic control of early embryonic development".

Plate I, Figure 3 in Velpeau (1833) illustrates one of the earliest stages where the axis of the embryo is horizontal; the drawing was taken from a work by the physician Gilbert Breschet (1784–1845). Plate II, Figure 1–1 illustrates a 20 day embryo that Velpeau had drawn as early as 1823 and which had "the head strongly bent forward". This is the cervical flexure. He corrected this by noting that the drawer took a bud of the arm for the bud of the mandible. It was possible to identify the stage where the human form was becoming more precise:

> After the sixth week, the organism of the embryo differentiates very quickly. The nose over all experiences remarkable changes. (ibid., author's translation)

Velpeau disproved the presence of gills. These illustrations did not give a description of the internal organization, such as of organs like the heart, or vascularization and cartilaginous outlines of the limbs and head, as the instruments did not allow this (histological sections to be mounted on the blade). However, they showed what could be compared between humans and other mammals and at what stage the macroscopic external appearance distinguished it from a sheep, for example:

> The human egg, like that of other mammals, consists of several membranes, a placenta, an umbilical cord and an amniotic fluid, known under the general title of appendages of the embryo or fetus (…). I do not know of anyone who has been able to deal with such a large body of facts (…). After consulting the annals of science, I saw that the great questions relating to organic evolution in the human species had hardly been decided except on evidence borrowed from analogy. Now, without wanting to contradict in any way the great principles already established, I believed that it was important to submit all these questions to counter-evidence, by descending from the human to other animate beings (…). I maintain that woman has to be taken in her turn as the starting point, before it is permissible to conclude rigorously, as a host of authors have too often tended to do until now. (ibid., author's translation)

Velpeau non-judgmentally evoked the issue of induced abortions and sympathized with the plight of women under duress, but he was critical of the violence of the methods.

Human embryology could only be seriously studied in a hospital and obstetrical environment. As long as the Chair of Comparative Embryogeny at the *Collège de France* was not associated with an obstetrics department of a Parisian hospital, the project of a natural history of human ontogeny in comparison to animals was doomed to failure.

9.3.3. *The first Chair of Embryogeny at the* Collège de France *(1844)*

In 1844, the *Collège de France* created the first Chair of Comparative Embryogeny for Victor Coste. Embryogenesis was recognized as being the establishment of large anatomical and skeletal structures and its consideration no longer supported Lamarck's post-natal transformism. Coste adhered to the school of Henri de Blainville, considering human embryogenesis as a creation distinct from that common to chimpanzees and the orangutan. The concept of an embryogenesis common to these two great apes was, moreover, something that had not yet been thought of, but which is still relevant today. The fossils of great apes and humans were unknown and the biology of gametogenesis did not allow for conceptualizing a transformist theory of embryogenesis. Human embryogenesis could therefore be reduced to the transformation of previous embryogenesis; for anatomists this concept was not rational and above all they found no reason for the need for such a transformation; the Supreme Being did not need the ape to create humans.

Coste was in charge of applied research programs to improve oyster production and other priorities. Fundamental research on humans and an evolutionary identity in their gametes was not a priority, neither from the point of view of knowledge, nor from the point of view of public health problems related to the verticality of the axial skeleton. French research lagged behind in the study of comparative embryogenesis. German physicians and anatomists, on the other hand, were particularly qualified and advanced, so the concepts were developed in German universities. As for primatology, it was not an autonomous science either, and great ape specimens were becoming rare.

9.3.4. *The discovery of the gorilla, 1847–1852*

In 1847, Thomas Savage (1804–1880), a missionary physician born in Connecticut (northwest of the Atlantic coast of the United States of America), brought back from Gabon four skulls – two males and two females – as well as bones that did not resemble the chimpanzee. He gave the first description to the Boston Society of Natural History in collaboration with a medical anthropologist, Jeffries Wyman (1814–1874), a Professor at Harvard University. He took courses in comparative anatomy and natural history in Paris and London in 1841–1842. He had to his credit a study on the chimpanzee or *Troglodytes niger* published in 1843–1844. The two doctors decided to name the animal *Troglodytes gorilla*. The discovery was far too extraordinary not to be verified by a confirmed primatologist like Isidore Geoffroy Saint-Hilaire. The existence of an ape different from the chimpanzee had long been suspected but no specimen had ever been reported in European collections. The opportunity arose in 1849:

> Both still relied only on the examination of the skull and teeth, as no one in Europe or America had been able to bring more complete elements of determination. These elements were finally partially obtained one year later, and it was the *Muséum d'Histoire naturelle de Paris* that received them. In April 1849, a surgeon of the French Navy, Mr. Gautier-Laboulaye, enriched the Museum, not only with an adult skull, but also with an equally adult skeleton, which immediately became the subject of much-followed studies by Mr. de Blainville. This was one of the works that our illustrious colleague was happily occupied with when he died suddenly. Such were the materials already possessed by the Museum and by science at the moment when, thanks to the double gift I have just announced, it became possible for us to be completely familiar with the gorilla in its

> organization, both external and internal, and to determine exactly its characters and its relations, both with the chimpanzee and with the orangutan. (Geoffroy Saint-Hilaire, *Sur le gorille*, 1852, author's translation)

Isidore Geoffroy noted enough different characteristics from the chimpanzee to conclude that it was a new genus that he named *Gorilla gorilla* in 1852.

9.3.5. *Franz Fick (1813–1858), a giant step forward: the study of the internal base of the skull (1853–1862)*

Despite the enrichment of the collections, the internal anatomy of the axial skeleton remained the *terra incognita* of primatology, *H. sapiens* included. The metric study of the internal surface of the cranial base was hampered by the difficulty of access due to the fragility of the bones. It required delicate cutting techniques to avoid destroying them. Studies began to follow one another and were completed during the second half of the 19th century with Fick (1853), Virchow (1857) and Welcker (1862).

Franz Ludwig Fick (1813–1858) was Professor of Anatomy at the University of Marburg (Central Germany). His research interests included the mechanisms of bone growth, particularly in the vertebrate skull (Fick 1853). His study was exceptional.

The metric comparison of the base was finally applied to a great ape, a non-primate mammal and a fish. For the first time, an anatomist looked at the combination of the brain and spinal cord in their adult bony architectural context.

Fick followed the direction of Aristotle's scale degradation: the upright human being, the semi-upright orangutan, a carnivorous mammal without flexion (a dog) and a fish (a pike), showing the base aligned with the spine without angular breaks. The study was entitled *Die Combination der Cerebralorgane mit dem Spinalsystem*, the skull drawings, of great finesse, show the figures in exocranial profile and in medial sagittal section. He also compared the skull of a human in lower exocranial view with that of a carp without flexion. Finally, he drew the compass used for his measurements and, most importantly, a diagram of the general organization of the spatial composition of the skull bones. This was the *Typen für die Form der Knochenbildung*, or *Morphotype of the ossified head*. It was no longer the comparison of the elongated skeletons of Étienne Geoffroy Saint-Hilaire.

This study was the first measurement of the angulation of the base, constructed with three homologous points: the basion, the back of the sella turcica and the point of contact between the frontal squama and the cribriform plate of the ethmoid. It also added gnathion, the lowest point of the mandibular symphysis (Figure 9.1).

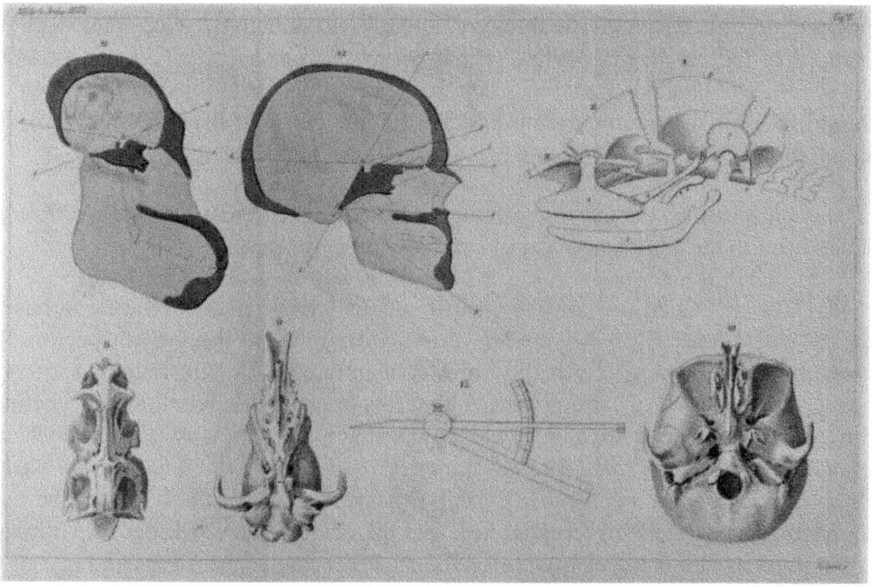

Figure 9.1. *At the top, on the left, mid-sagittal sections of orangutan and human skulls, showing for the first time the angle between the clivus and the floor of the frontal lobes; on the right, a diagram of the general organization of an adult skull with the base flat, the mandible long and low. Below, on the left, a long and narrow pike skull in superior and inferior views, its base is flat; on the right, the skull of an adult human in inferior view with the occipital foramen clearly visible (Fick 1853)*

The sections clearly show the retraction of the two dental arches of humans under the nasal fossae, the protruding chin, the straightening of the clivus (the posterior sphenoid and the basi-occipital aligned) and the widening in the transverse plane with the separation of the two petrous pyramids pushed forward, but without angular quantification.

Fick thus showed that the structure of the skull base and the face of the human is organized in the three planes of space around the angle of flexion. According to his diagram, the plane conformation was the archetype; the great ape's conformation was already far from it and the most verticalized was the that of the human being, with the least prognathic face slipped under the frontal lobes.

9.3.6. *Rudolf Virchow (1821–1902), a major study on the relations between the internal base and the external face*

Rudolf Virchow (1821–1902) was Professor of Pathological Medicine at the University of Würzburg (Southern Central Germany). In 1855, he formulated a principle on the basis of the biology already described by von Baer with the divisions of the fertilized oocyte: *every cell comes from another cell* (Heinz 1991). In 1857, he published a major study on the angular relationships between the internal base of the neurocranium and the middle stage of the face in the median sagittal plane (Virchow 1857).

This publication was the first monograph showing different fetal stages from the human head to the adult stage and illustrated with color plates.

Virchow described the bony and cartilaginous territories of the face and base as well as the famous *spheno-occipital synchondrosis* where the growth vectors that drive the base toward verticality are concentrated. An illustration shows an anencephalic fetus with strong verticality, the brain and cerebellum degenerated, with only the brain stem and spinal cord remaining. Virchow echoed Fick and defined several angles "*to locate the bones of the face and the bones of the base of the skull in relation to each other*", front to back: the root of the bones of the nose, the maxilla, the clivus, the occipital hole and the occipital scale (Table V in Virchow (1857, p. 63)). These angles quantified the straightening of the clivus and that of the midface.

At this stage, prognathism is not influenced by the development of teeth and occlusion, and its relationship with the flexion angle is understood before birth and mastication. *The author did not situate the flexed conformation of the human base in the continuity of embryogenesis because it was still unknown*. Nevertheless, this study could now be used as a metric protocol for comparative morphogenesis of primates as collections became available with skulls at different stages after fetal growth. The base angle would vary between authors.

It is easy to understand why Virchow did not adhere to Lamarck's transformism and shared the position of Cuvier and de Blainville. No locomotor function could be responsible for the fetal growth of the skull base; the organ correlated to it is the brain and the continuum between the brain stem and the spinal cord.

Nothing justified a morphogenic reorganization of the ape into the human, and therefore no rational position could justify the need for an intermediate and disappeared morphotype between that of the great ape and *H. sapiens*.

9.3.7. *Neanderthal Man (1856), a lost human species*

An incomplete skeleton had just been discovered in Germany by two workers in a small cave near Düsseldorf in 1856. They handed over the bones to a teacher from the neighboring town, Johann Carl Fuhlrott. The differences from humans were significant. He therefore consulted an expert in archeological anthropology, Hermann Schaaffhausen (1816–1893), who taught at the University of Bonn. As early as 1857, this anthropologist confirmed his conclusions: this skeleton had an unknown anatomy, it was extinct, and he declared himself in favor of an animal origin for humans in a descent of pithecoid animals via gradual evolution. The fossil was a lost species in the 18th century sense.

With regard to the skull, only the calvaria (cap) remained, but the whole showed a morphology without possible confusion with any pathology of *H. sapiens* (Schaaffhausen 1858). The skeleton included the calvaria, the two femurs, the right scapula and a clavicle, the bones of the arms, a vertebra, ribs, the left pelvis and the two humeri. The bone structure was organized around the axial verticality as shown by the conformation of the femurs, the coxo-femoral joint and the morphological structure of the shoulder. The skeleton was that of a biped similar to humans and showed no trace of arboreal function. The shape of the brain visible on the endocranial face was that of *H. sapiens*. While the bones were homologous and arranged along the vertical axis, the shapes differed. Less cerebralized, he called this species "Neanderthal Man", the fossil man of the Neander valley. Rudolf Virchow did not accept this diagnosis and interpreted the entire skeleton as a pathology unknown to the *sapiens* species. Perhaps he would have adopted a different attitude if he had the base of the skull and the entire face with the mandible. However, it had to be acknowledged that comparison of the basicranial growth between apes and humans had not yet been the subject of a study.

9.3.8. *Herman Welcker (1822–1897), comparative internal growth of the orangutan and* Homo sapiens *(1862)*

In 1862, Herman Welcker (1822–1897) published the first comparison of the growth of the sphenoidal angle and prognathism between humans and orangutans. The orangutans were studied with three specimens of increasing age after birth, while humans were studied from the neonatal stage. The sphenoidal angle was measured by taking an exocranial facial point, the nasion. Fick's protocol (1853) was preferable because the growth of the face varies with the development of the teeth. The homologous points showed the opening of the angle in apes with increasing sub-nasal prognathism (Taf. X, Fig. 1 and Fig. 2 in Welcker (1862), available online). Thus, not only was the straightening angle less pronounced in the young orangutan, but the clivus lost its verticality with age. The growth vectors of the face and base did not tend toward verticality, but toward horizontality; in other words the posture returned to a more stable quadrupedal balance. This comparison did not bring any argument to Lamarck's theory; on the contrary, the morphogenesis of the ape moved away from bipedality, while the young moved in bipedal equilibrium. The anatomist did not specify this *but it was the masterly demonstration that invalidated the primary bipedal locomotion in the development of the ape as the cause of the phylogenetic axial straightening*. The use of bipedality decreases with the end of neural growth.

10

The Decline of Transformism at the *Muséum national d'Histoire naturelle*

10.1. The decline of transformism despite the discovery of the first monkey fossils

Transformism was not supported other than by Étienne and Isidore Geoffroy Saint-Hilaire in the Museum of Paris. Fossils helped a little but did not lead in the expected direction. In 1834, Edouard Lartet (1801–1871), a lawyer and amateur paleontologist, had contacted the father Étienne Geoffroy to inform him of a fossiliferous deposit located at Sansan in the Gers (southwestern France) dating from the Tertiary era. In 1836, he announced the collection of a monkey mandible resembling that of the present-day gibbon but known in Southeast Asia. The fossil was studied by Henri de Blainville, who named it *Pliopithecus*. It dated from the Miocene, about 15 million years ago. The discovery caused a sensation because it proved Geoffroy right against Cuvier's doctrine. In 1841, Isidore Geoffroy succeeded his blind father and found himself in an academic position to defend the theory of the simian origins of humans. Étienne died in 1844 and Isidore was then alone in Europe in maintaining the Lamarck school; the Museum of Paris was still the very fragile intellectual crucible of the *Homo sapiens*'s simian origin. Born at the *Jardin des plantes* in the pavilion attributed to his family, Isidore had been his father's naturalist assistant from the age of 19 years. He was 24 years old when Lamarck died in his own pavilion, also located in the *Jardin*. He therefore had time to get to know him. Isidore was elected member of the *Académie des sciences* in 1833 (anatomy and zoology section). At the death of his father, his knowledge of the comparative anatomy of current and fossil primates, as well as human fetal teratology, gave him a field of knowledge and an authority unique in Europe. He could claim to support evolution as the primary principle of the unity of extinct and present-day organisms, while theologians firmly believed in the creation of humans 9,000 years earlier and therefore no primate fossils would change the doctrine.

For stratigraphic geologists, these 9,000 years correspond to the post-Diluvium age, that is to say the last stage of the Quaternary era, or the Holocene. Thus, in 1841, the order of monkeys had officially existed since at least the Tertiary era, and the order of humans did not exist before the Holocene; but a 15 million year hiatus separates fossils and humans. New archeological findings supported the paleontological collections by showing the presence of human activities in the river terraces of the Diluvium. They were not supposed to have existed.

10.1.1. *First manmade objects contemporary to Diluvium, 1842*

Jacques Boucher de Perthes (1788–1868), a lover of botany and archeology, prospected the natural terraces of the Somme (Picardy, northern France) and collected flints of beautiful workmanship arranged on two sides. In 1842, he collected one of these bifacial cut stones in a very old terrace, associated with a mammoth mandible. A corresponding member of the *Académie des sciences* since 1800, he was appreciated for his scientific qualities. He therefore informed the Academy, but no member followed up to verify the facts. Was Pope Pius IX concerned about this? If humans dated from the Diluvium, and if an order close to humans had existed before Diluvium, the Bible would be nothing more than a book of imperfect knowledge and therefore a collective work written by mortal humans. In 1847, Pius XI therefore created the new Pontifical Academy of Sciences and declared it the only one able to deal with "human facts": medicinal, religious and historical, philosophical, moral and juridical sciences were to remain under the control of the Holy Office. The "human sciences" could not exist in themselves; the origins of humans were a matter of religious doctrine. Only Catholic scholars were concerned, and if so only because it was important for them not to upset the scholastics.

Across the Channel, several scientists of European renown, including paleontologist Hugh Falconer (1808–1865), geologist Joseph Prestwich (1812–1896) and archeologist-geologist John Evans (1823–1908), made the same observations as Boucher de Perthes. Lithic tools were collected in the karst deposits of the Brixham Cave in Devonshire (southwest England). They did not have to account for their reflections to the Holy Office and it did not matter to them what the members of the Academy of Sciences of Paris thought. In 1859, they went with Boucher de Perthes to the site and validated his observations. The geologist Charles Lyell, considered the European authority in stratigraphic geology, announced all the British and French discoveries to members of the British Association for the Advancement of Science. Boucher de Perthes was recognized by the English as the founder of prehistory, or the science of human productions that preceded history. This was not very flattering for the members of the *Institut de France* and the professors of the Museum, who could have made the trip, 100 kilometers from Paris, while the British were crossing the English Channel.

Two notable professors from the Museum of Paris visited the site, the paleontologist Albert Gaudry and the naturalist physician Armand de Quatrefages. Their excavations confirmed the stratigraphic authenticity of the tools and fossils. The Academy officialized the existence of a human intelligence dating from the Diluvium, thus doing justice to Boucher de Perthes. Cuvier's dogma of the succession of creations subordinate to the great extinctions and revolutions of the globe was being challenged.

Armand de Quatrefages (1810–1892) had been a member of the *Académie des sciences* since 1852 and held the Chair of Anthropology and Ethnography at the Museum of Natural History since 1855. He combined his skills, first as a physicist and mathematician with a thesis on ballistics and a theoretical study on lunar volcano projectiles (the question of gravitation); he was also a doctor with a second thesis directed by Georges Louis Duvernoy. Finally, de Quatrefages was an anatomist with a third thesis on rodent teeth.

Following the death of Duvernoy in 1855, who held two professorships, there was a game of musical chairs between the Museum and the *Collège de France*, closely linked since the creation of the *Jardin royal des plantes*. As we have seen, Étienne Serres succeeded him to the Chair of Comparative Anatomy at the Museum and left his Chair of Anthropology to Armand de Quatrefages, a monogenist like the German Blumenbach. Cuvier's chair at the *Collège de France* was transferred to a second successor, Pierre Flourens. None of these professors adhered to Lamarck's transformism. Paradoxically, the *coup de grâce* was brought by Edouard Lartet who discovered the fossilized small ape *Propliopithecus*.

10.1.2. *Edouard Lartet: the first great fossil monkey to the rescue of Cuvier (1856)*

Edouard Lartet became Geoffroy's assistant in 1851. The year 1856 remained the year of consecration, with the discovery of a half mandible of a great ape dated 12–10 million years ago. The fossil had been collected by Dr. Alfred Fontan in a paleontological site at the commune of Saint-Gaudens in Haute-Garonne (southwestern France). The mandible had characteristics common to chimpanzees, gorillas and orangutans. Lartet named the fossil *Dryopithecus fontani* or "oaks ape" in reference to fossil woods found in equivalent fossiliferous levels in the Aquitaine Basin. The Museum's collections now included two cranial monkey fossils, one whose ancestral lineage was that of current gibbons, the other whose lineage was that of African chimpanzees and gorillas and the Asian orangutan. The cast of *Paidopithex*'s femur sent to Cuvier was probably not in the collections, but three fossil specimens now proved that tree-dwelling apes, as diverse as a type of gibbon and chimpanzee, lived in a tropical climate in southern and central Western Europe between 15 and 10 million years ago. The temporal hiatus with humans was shrinking.

For Lartet, however, the fossil did not justify transformism; on the contrary, the mandible confirmed the stability of the axial structure with a variability of the growth rates of the face. The appendicular skeleton of the three African and Asian genera simply attested to different arboreal and terrestrial locomotor adaptations of the same axial structure. Edouard Lartet joined de Blainville and Victor Coste: humans were neither a species, nor a genus, nor a family of monkeys, but an order in their own right. The year 1856 was not to be outdone for human paleontology either, with the discovery of the Neanderthal skeleton, but Schaaffhausen's work published in 1858 left many paleontologists and anthropologists in expectation or disbelief.

10.1.3. *Paris, capital of transformist anthropology and free thought (1848–1857)*

The second half of the 19th century brought its share of setbacks to the biblical doctrine of the creation of humans. Paul Broca (1824–1880), a French doctor attached to the Faculty of Medicine at the Sorbonne (Paris), of Protestant but agnostic faith, was determined to put an end to the dogma of monogenism. In 1848, a few weeks after the third French revolution and the famous papal bull of 1847, Paul Broca created the *Société des Libres Penseurs* (Society of Free Thinkers), the headquarters of which were at the Faculty of Medicine and which was driven by a spirit of intellectual autonomy and sovereignty:

> If scholars are passionate, more or less, depending on their temperament, about philosophical or religious, social or humanitarian problems, nothing could be better. But when they return to their laboratories and apply themselves to scientific research, they must compress their feelings and aspirations and close their ears to outside noise, to hear only the inflexible voice of truth. *For science must depend on itself alone, and cannot bend to the demands of the parties; it is the august goddess who reigns above mankind, to direct it, not to follow it, and it is of her alone that it can be said that she is made to command and not to obey.* (Broca, *Histoire des progrès des études anthropologiques depuis la fondation de la Société*, 1870, author's translation)

In 1857, as teaching was free, Broca set up anthropology courses that he taught in the amphitheaters and dissection pavilions of the Paris Faculty of Medicine. In 1859, he founded the Anthropological Society of Paris (*Société d'Anthropologie de Paris*) with the support of Isidore Geoffroy Saint-Hilaire, who, in 1854, had created the Imperial Zoological Acclimatization Society. It was a learned society whose objectives were the introduction and acclimatization of exotic animal species in France, not only for agricultural and commercial purposes, but also as cultural education for the general public. The park was located in the *Bois de Boulogne*, west of Paris and near the Seine River. In the same year, 1859, the society took the name *Jardin d'Acclimatation*, a garden which would henceforth be closely linked to the development of the Anthropological Society of Paris.

Paul Broca had shared with Isidore Geoffroy Saint-Hilaire an investigation into hybridization between a rabbit and a hare, despite their being two different genera. He verified whether the hybrids were fertile by crossing them and the results were conclusive. The barrier between genera allowed for postulating the impossibility of modifying them gradually by hybridization. This investigation invalidated fixism. Paul Broca was therefore not a monogenist and he found in this experiment a demonstration meaning that human crossing breeds were not proof of a single couple but of several crossings with a polycentric drift, as a consequence of which a single strain of a fossil genus was not credible.

The doctor had tried to present this study on rabbit hybridization to the *Société de Biologie* (Paris) but was censored when he extrapolated his analogies to human crossbreeding. This censorship was at the origin of the Anthropological Society of Paris, with the support of Isidore Geoffroy Saint-Hilaire, the only evolutionary paleo-primatologist, who was close to Lamarck's ideas and had known him. Professor at the Museum, holder of the Chair of Zoology and member of the *Académie des sciences*, he was an authority in comparative anatomy, but faced the power of the Catholic Church, which did not admit the simian origin of *H. sapiens*. The founders of the two learned societies thus benefitted from federating their knowledge.

The doctor Broca published his essays on hybridization as early as 1858 (Broca 1858–1859, 1859–1860), as well as a reflection on the "primitive races" of Diluvium as early as 1859. In 1867, for his personal research, he created the Anthropology Laboratory at the *École Pratique des Hautes Études* (EPHE) hosted by the Faculty of Medicine; then, in 1872, he founded the *Bulletins d'Anthropologie*; finally, in 1876, he founded the Paris School of Anthropology (*École d'Anthropologie de Paris*).

The Society aimed to be the center of reflections on the origins of humans, involving leading personalities who had good anatomical knowledges of current and fossilized apes, such as Isidore Geoffroy and Edouard Lartetand, and knowledge of human physiology with physicians such as Louis Pierre Gratiolet (1815–1865; neurology) (Gratiolet 1854), Paul Topinard (1830–1911) and Jacques Bertillon (1851–1922) (demographer and statistician). Armand de Quatrefages represented the authority of the National Museum of Natural History and the Academy of Sciences in anthropology. The Society established itself as the European leader in anthropometry and the classification of human races through the use of craniometry. The first question was therefore that of monogenism or polygenism. The notion of race dominated European anthropology and it was sufficient that the idea of a graduated evolution be accepted by an influential learned society to see its extrapolation to human races, now seen as so many grades of the past that would still persist in the 19th century:

> Since the existence of paleontological man has been demonstrated, the position of monogenists has become more difficult. They find it easier, it is true, to explain the formation of types, but they cannot do so without invoking principles whose application would tend towards transformism. (Broca, *Sur les races primitives contemporaines de l'époque dite du Diluvium*, 1859, author's translation)

Armand de Quatrefages was a monogenist and non-evolutionist; he reminded Paul Broca of the counterproductive effects of hybridization between genera, well-known to sheep breeders and botanists. Certainly, these cross-breeds showed the conservation of hybrid traits for a few generations, but these transient variations disappeared with a return to the phenotype of the two strains. This observation was called "the law of reversion". Moreover, this "law" invalidated the Lamarckian hypothesis of a possible gradual transformation at the level of the higher anatomical organization that defined the genus to which the principle of correlations and organo-skeletal coordination applied. Paleontologists were aware that the system could not reorganize itself without simultaneous coordination of the parts, yet the corpus of scientific knowledge was far too insufficient to construct a theoretically demonstrative mathematical synthesis. The debates thus show the general confusion in which scholars struggled to make concepts explicit and clear, involving Isidore Geoffroy Saint-Hilaire, the physicians and Paul Broca, who was looking for an alternative to the transformation of organizational plans.

10.2. A theory lacking internal coherence

10.2.1. *Gradualist classification and discontinuities between fossil genera: an impasse*

Paul Broca accepted Neanderthal Man as early as 1857. This raised the question of how Neanderthals should be classified. How should they be situated in natural history? Was the Neanderthal a disappeared order? Since Linnaeus, *Homo* had been synonymous with *sapiens*, it was one and the same organization. For Cuvier and de Quatrefages, it was an order in its own right, the bimane, and there was no primate order as Linnaeus maintained. Important discoveries in paleoanthropology and prehistoric archeology would soon change the state of the art.

In 1863, Edouard Lartet undertook excavations in caves in the Périgord with the banker and ethnologist Henry Christy (1810–1865) and discovered at Le Moustier site an assemblage of lithic tools more elaborate than the bifaces of the Somme. The prehistorian Gabriel de Mortillet (1821–1898) named this techno-typology of cut stone the Mousterian, which would become the emblematic culture of the Neanderthals.

In 1863, Irish geologist William King (1809–1886) proposed to the British Association for the Advancement of Science that the Neanderthal skeleton be named *Homo neanderthalensis*, estimating its age at 30,000 years. In 1864, the British Association recognized the skull of Gibraltar as a Neanderthal; in 1866, a mandible was discovered in Belgium in the cave of La Naulette (Province of Namur), by the geologist Édouard-François Dupont (1841–1911). The bone was associated with fossils of mammoths, rhinos and reindeer. Dupont described it and had it drawn; the jaw differed from *H. sapiens* (Dupont 1866). Finally, in 1868, Edouard Lartet and Henry Christy exhumed five *H. sapiens* skeletons in the Cro-Magnon rock shelter at Les Eyzies-de-Tayac (Dordogne). The excavations showed a sepulchral context with ornaments (shell necklaces), as well as elaborate and diversified lithic tools dispersed in the soils of occupation. These were the oldest representatives of *H. sapiens*, dated to 27,000 years ago. The dates were three times older than those accepted by the church. Funeral rites and artistic and symbolic thoughts were already well-constituted and did not reflect Buffon's image of a semi-wild humanity terrified by the volcanic eruptions of a last phase of the cooling Earth. As for Genesis, the theologians were forced to acknowledge that it was written by human hands.

The British Association for the Advancement of Sciences had validated William King's Linnaean nomenclature *Homo neanderthalensis*. However, anatomists did not classify a skeleton before having measured it and compared it to a significant sample of fossil and current species. Creating a fossil species of the genus *Homo* was not conceivable until a study showed a level of morpho-skeletal *Homo* organization common to the fossil and to present-day humans, this organization being the oldest, while *sapiens* would be only a variety. However, was this not a way of getting rid of the fixism or evolution problematic, by presupposing a single plan of morphogenetic organization?

The base of the skull was missing, the angle of straightening (or sphenoidal angle) was not measurable, but the high position of the occipital bone indicated that the brain was less coiled down and forward and, therefore, the clivus was less straightened. It also meant a more significant facial prognathism, and this was indeed what La Naulette's mandible showed, with its chinless and receding symphysis. Paul Broca wondered whether, with such a non-human craniofacial morphology, the Neanderthal was gifted with articulate language. The question interested him all the more since he demonstrated in 1861 the location of the cerebral area of this function in the left lower frontal cortex, which monkeys lack.

There did not appear to be any morphogenetic uniqueness for *H. sapiens* and Neanderthal Man. The *sapiens* anatomy, which had also defined the genus *Homo* since Linnaeus, was not a recent variation of an ancestral morphology (and more rigorously of a morphogenesis) common with Neanderthals. The nomenclature criteria for defining the genus were therefore not respected, while the scarcity of paleontological data showed that the Neanderthal posed a problem for nomenclature. The place of humans in the classification, and their natural relations with apes for those who believed in them, became a subject of discussion of primary importance, because, if present-day humanity was the result of the evolution of primates such as *Dryopithecus*, nothing would prevent the evolution of human from continuing, which was even more unthinkable to theologians, since for them, creation finished with the Adam and Eve model and its descendants. In a context where evolutionism was beginning to be emulated with the hybridization of Leporidae as its only model, the transposition to human beings was becoming a very worrying subject for society, not for understanding the past, but for the future of humans and the status of each human being. Armand de Quatrefages was a monogenist and, from his point of view, hybridization did not change the biological unity of humans and had no evolutionary impact since he did not believe in it. However, it was different for the followers of the emerging doctrine of eugenics, or the improvement of the human race. This was how an evolutionary anthropology began to germinate and feed the eugenicist theories supported by a so-called "driving force" of evolution.

10.2.2. *Charles Darwin (1809–1882) knocked at the door of the French Academy of Sciences and convinced no one*

Isidore Geoffroy Saint-Hilaire, Armand de Quatrefages and Pierre Flourens received letters from a British naturalist who was interested in Lamarck's theory and the debates on hybridization. This naturalist sent them his work published in 1859 and entitled *On the Origin of Species by Means of Natural Selection, or the Preservation of Favored Races in the Struggle for Life*. Armand de Quatrefages knew the sender well: the author was Charles Darwin, with whom he had corresponded for several years. The archives of the British naturalist preserve 19 of these letters exchanged between 1855 and 1873.

Charles Darwin had been generating controversy in his own country since paleontologist Thomas Huxley (1825–1895) commented on his work before the British Association for the Advancement of Science in 1860, encountering strong opposition from paleontologist Richard Owen (1804–1892), who did not validate the scientificity of his reasoning. Richard Owen had been Professor of Anatomy and Physiology at the Royal College of Surgeons since 1836. He had known Georges Cuvier since the latter had visited the College in 1830. In return, Owen was invited to meet him in Paris in August 1831. The conflict with Geoffroy Saint-Hilaire was then at its height. The two British and French luminaries took the time to discuss the questions of homology and kinship between the two invertebrate and vertebrate branches necessary for the transformist thesis of a single composition plan. Later, in 1848, Owen analyzed the consequences of von Baer's law and invalidated it from his point of view, finding counter-examples (Owen 1848). Von Baer's law stipulates a linear evolutionary mode from fertilization, integrating, in the chronological order of the development of the embryo, the major stages of the evolution of types from the most primitive.

What Owen and Cuvier criticized about the same agreement was the immediate acceptance of the possibility of moving naturally from a simple organization to a more complex integration from embryogenesis onward. The criticism was given by the observation of characteristics common to invertebrates and vertebrates but which appear late in their development.

However, for a transformist nomenclator like Geoffroy Saint-Hilaire, vertebrates appeared millions of years after invertebrates. For Owen, these characters therefore did not signify a transformation of the invertebrate structural plan into a vertebrate structural one but were independent creations. The paleontologist interpreted the animal kingdom as the coexistence or succession of several independent embryonic organization plans, which he called "archetypes" (Gaudant 1992) and from which new species will bud, becoming the strains of new buds. However, none of these buds can generate a restructured and more complex organizational plan than the primitive strain of which it kept the fundamental structure:

> Darwin relates the meaning of the primitive plan represented by the archetype to that of a nucleus of a system of adaptive variations whose organizer is natural selection; Darwin used unity of plan as an additional argument to convince his audience of a kinship between all species of the same type. But homology had encountered contradictions: vertebrate organs were found in invertebrates, which undermined the *principle (law of differentiation of drafts) of the ontogenetic precedence of von Baer's homologies*. (Balan, *L'ordre et le temps*, 1979, author's translation)

Charles Darwin believed that he had gathered a sufficient amount of knowledge to overcome the difficulty raised by Cuvier, thanks to his bibliographical erudition and personal observations, enriched by his expeditions to the distant colonies of the Victorian Empire. He was a Lamarckian, convinced of the hereditary transmission of the characteristics acquired, supposedly, during the life of the animal. He tried to unify the whole, like Jean-Baptiste Lamarck, while taking up his powerful concept of circumstances favorable to the hereditary transmission of acquired traits. Charles Darwin considered his demonstration a great innovation because he would have discovered the "engine" driving evolution. This "engine" would be "natural selection in the struggle for existence". However, the term "engine" is only a metaphor because Charles Darwin was not interested in the physico-chemical processes enunciated by Lamarck in his 1794 essay. Darwin did not address the impasse on the mechanisms of transformation of forms. In an attempt to win an audience, he argued that Lamarck did not insist enough on environmental pressure to justify the transmission of transformations.

The basis of his reasoning was built on the idea that the population was made up of unfit people – that is to say, those who were said to be untransformed by the environment – and who were therefore eliminated naturally. This reasoning became a tautology: transformations were transmitted only by the transformed or the "better adapted"; the non-transformed or the maladjusted died.

Did not Darwin perceive that, for paleontologists like Owen, he was dodging the real problem, namely, the process of reorganizing morphogenesis? It was obvious, for an evolutionist, that modifications were transmitted by those who would have been transformed by the environment, if it were "materially" possible. His model was a circular reasoning that did not deal with the process of transformation. This was where the problem lay: naturalists expected a true mechanistic explanation with references to the laws of physics and chemistry.

It was the process of organic transformation that needed to be explained, along with its transmission to gametes. The obstacle that Darwin did not solve concerned the level of organization transformed. If, according to him, the change began with the post-natal structure modified by the function of the associated organs, how were these changes transmitted to support heredity?

The term "natural selection" did not refer to a physiological mechanism, it was an expression to qualify the success of transmission by maintaining favorable circumstances, which was Lamarck's principle.

The only truly innovative idea was the relationship between the adapted and the maladjusted, who presented a risk of extinction, but extinction by maladjustment was not a new idea; it was Cuvier's catastrophism.

In other words, Darwin lacked what physical and chemical naturalists expected from an explanation of the transformations. To be credible the transformist needed to know how to describe the forces at work at the molecular scales of muscles, organs, the skeleton and joints and explain the transmission to the gametes.

The book therefore did not find a scholarly audience in France for these legitimate scientific reasons and not for creationist reasons. Paul Broca did not adhere to transformism but he was nonetheless an evolutionist in the sense that he accepted this past reality, without the mechanisms being demonstrated for all that. The physician and leader of free thought recognized the necessity of natural links between fossil and living species in order to justify the anatomical proximities of skeletons, but he did not adhere to Lamarck's speculation and even less to the explanatory power of natural selection:

> Natural selection is in contradiction with the facts, since its fundamental principle is to be slowly progressive. For it must be recognized that in certain species, the orangutan type, for example, there has been transfiguration completely carried out in one go, a true supernatural act, equivalent to an act of creation. (Broca in Huard, 1961, author's translation)

He developed regular correspondence with Isidore Geoffroy, who found no reason to evoke a scientific discovery or an original conceptualization. Indeed, Charles Darwin took up the two innovative theories, that of Étienne Geoffroy on the possibility of a gradual transformation within the limits of a single unity of composition plan and, from Lamarck, that of acquired characteristics, transmitted by heredity as soon as ecological conditions became favorable.

It was a digest of these concepts that the naturalists of the Museum of Natural History in Paris had been arguing over since the conflict between the Lamarck–Geoffroy Saint-Hilaire tandem and the Cuvier–Owen tandem, made worse by the disagreements between de Quatrefages and Broca on the capacity of the hybridization of genera to generate a new genus.

The absence of knowledge of the laws of physics to which organisms are also subject, and which refer to dynamic systems coordinated between organs and structures, did not allow for adherence to Lamarcko-Darwinian logic. The mechanisms that were needed to account for the simian origins of human organization were misunderstood and constitute a considerable field of research that is still relevant today.

The French naturalists who received the missives and the book by Charles Darwin, therefore, learned nothing that had not already been discussed for more than half a century between the professors of the *Collège de France* and the Museum and within the *Académie des sciences*. These men of science were waiting for a scientific theory that would explain how a morphogenic reorganization could emerge from fertilized eggs and modify the embryonic leaflets described by Wolff.

From the first weeks of development, the theoretical problem was the science of molecular or corpuscular dynamic systems, which consumed energy. Morphogenesis was concerned by the principles of thermodynamics. It was still out of the question, whereas these considerations were primary, as Sadi Carnot understood it, who died prematurely in 1832, when the need to recognize the evolution of organizational plans and the understanding of its principles was precisely raised.

The title *Origin of Species* did not therefore correspond to a discovery of physico-chemical mechanisms making it possible to make inferences from laws or principles of organization and manufacture molecules from fertilization. French naturalists, evolutionists or not, were still hungry.

As for the origins of humans, Charles Darwin addressed them all the more since the classification of primates deserved to be revisited due to the large number of species discovered in America, Africa and Asia. In 1863, Thomas Huxley published a nomenclature of primates entitled *Evidence as to Man's Place in Nature* in which he recognized Linnaeus' classification with a single order, that of primates divided into five families including humans. Axial straightening remained the blank on the map. Armand de Quatrefages did not change his positions; *H. sapiens* formed an order on its own.

10.2.3. *A progression toward scientific formalization of the evolution of structures: the geometrical study of the skull*

Following Welcker's work published that same year, 1863, Paul Broca invented an instrument to measure the sphenoidal angle without destroying the skull and presented it in 1865 to the *Société d'Anthropologie de Paris*:

> The sphenoidal angle, according to Mr. Welcker, would be at its maximum in the newborn child; it would then gradually decrease and go down to its minimum in the fully developed skull (…), it would also have the advantage of providing a very important distinctive character between man and ape. It would diminish in man from birth to the end of growth, while in monkeys it would become increasingly obtuse, and would even end up, in adulthood, becoming equal to two right angles. The importance of these conclusions will not escape anyone's notice. (Broca, "Sur un moyen de mesurer l'angle sphénoïdal sans ouvrir le crâne", 1865, author's translation)

The angle became more closed in humans as soon as the fetus grew, and after birth it closed, whereas in all apes, small or great, not only was it more open in the young (no-one was yet familiar with monkey fetuses) but it also opened up more with post-natal growth. For prosimians the question did not arise, since the angle did not exist:

> It is clear that, in order to take full advantage of the measurement of the sphenoidal angle, it is essential to study a very large number of skulls from this point of view. Up to now, this measurement could only be done on skulls that had been cut, which considerably restricted the number of cases where it could be done. Mr. Virchow measured the sphenoidal angle on skulls subjected to vertical and median cutting (…). The procedure I have just submitted to you allows us to construct the NSB triangle (nasion, sphenoid, basion), which I will call the sphenoidal triangle, without making any cut and without the slightest chance of error. (ibid., author's translation)

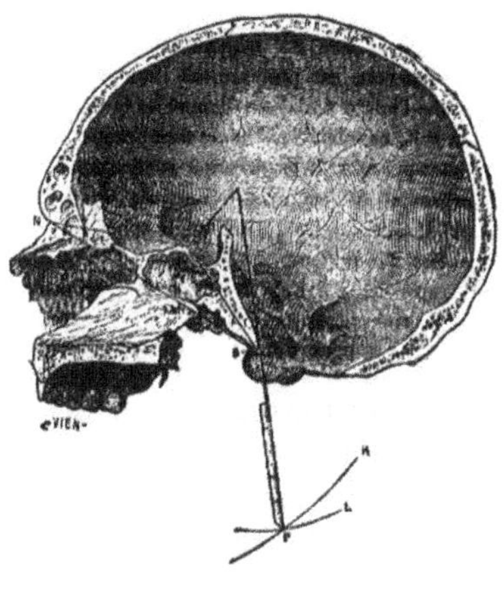

Figure 10.1. *The goniometer for measuring the sphenoidal angle. Broca, "Sur un moyen de mesurer l'angle sphénoïdal sans ouvrir le crâne", 1865. © UMR 7194 CNRS – Bibliothèque de l'Institut de paléontologie humaine*

Tensions were high within the Anthropological Society between Lamarcko-darwinians and structuralists. In 1869, Paul Broca organized a session to discuss this (Broca 1869). The French translation of Charles Darwin's book was presented by Eugène Dally (1833–1887), a medical psychotherapist sympathetic to transformism, who defended a dubious hypothesis on the proximity of the brains of criminals and apes. However, he did not master the comparative anatomy of primates. The anthropologist and ophthalmologist Franz Prüner-Bey (1808–1882), who shared the arguments and critical sense of Armand de Quatrefages, challenged him "to prove that there are fewer differences between man and certain apes, than there are between them and other apes". The challenge consisted of demonstrating the authenticity of a structural organization common to the great apes (chimpanzee, orangutan and gorilla) and to humans, which distanced them from the monkeys, in order to justify the break-up of the quadrumanes into parallel families. The vertical structure "human" no longer occupied the rank of an order; it was at the same level as any ape and in the same "box" labeled "order of primates".

Dally was not able to argue on an anatomical basis. His reasoning consisted of modifying the classification to make it compatible with his presupposition of a gradual evolution from monkey to human. Paul Broca was annoyed:

> Instead of dealing only with the question on the agenda, he also dealt with the question of transformism (…). In this way the two questions, although essentially distinct in principle, were in fact merged. I very much regret this (…). Today, therefore, I will limit myself exclusively to seeking what place comparative anatomy should assign to the human genus in the classification of mammals. (Broca, "L'ordre des primates, parallèle anatomique de l'homme et des singes", 1869)

10.2.4. *The cart before the horse: the Linnaean classification of fossils before understanding of processes*

It is important to understand the issues at stake. Dally followed the simplifying, if not simplistic, linear reasoning of Linnaeus' classification, which was thought up at a time when evolution was not the natural dynamic that accounted for the kinship links between present and fossil species. The principle of the classification, as we have already mentioned, was that of Russian doll nesting. The choice of a Linnaean taxonomy biased from the outset the understanding of transformation processes.

Huxley's reasoning followed the linear logic of the accumulation of traits that should have been the mechanism from the embryogenesis of the primate population strain. However, it was at the root of the evolution of embryogenesis that the inadequacy between the Linnaean classification and the processes of transition/transformation from prosimian to H. sapiens *arose.*

Homo sapiens was not the last diversity of an embryogenesis at least 60 million years old. Paul Broca, Armand de Quatrefages and Franz Pruner-Bey, among others, were aware of this reality, because the mandibles of the *Dryopithecus*, chimpanzee, gorilla and orangutan have the same structure, unchanged for at least 12 million years, and it has nothing in common with humans. This structure is long and

U-shaped and the symphysis is receding, whereas that of humans is parabolic, short and its symphysis is inverted, the chin protruding with the two edges of the lower part disjointed. On the other hand, at the level of the face, the mandible is dependent on the maxilla with which it articulates through dental occlusion and which has nothing comparable with humans; at the level of skull base, the mandible articulates precisely with the territory which is straightened and which shows opposite vectors of growth between humans (toward verticality) and apes (toward horizontality).

The level of morphogenetic organization that would classify great apes and humans in the same category did not exist. At no stage of morphogenesis did the human being superimpose its organization on that of a great ape, nor vice versa. Classifying them in the same category could only obscure understanding of the transformation processes.

The embryogenesis of H. sapiens *was not the Linnaean classification, or the phylogenesis of adult forms. The morphogenetic discontinuity with the great apes was objective. There was no morphogenetic organization plan common to prosimians, monkeys, great apes and humans.*

Broca therefore took up the question again and published his nomenclature of primates in 1869 after having reviewed a very large number of osteological (morphology and angulation of long bones, joints, suture fusion speed, etc.), muscular (numbers, shapes and insertions), organic (the brain, liver, reproductive organs in males and females and blood circulation) and physiological (length of gestation, highly variable placenta, digestive system, respiration, phonation, vision and urogenital system) data. He retained the order of primates and the division into five families starting with the human family. The criteria corresponded to the adult stage, external and in the appearance of posture and locomotion.

The internal organization of the base of the skull and the spine, even reduced to its medullary canal, were not taken into consideration. The bony and membranous envelopes surrounding the brain, cerebellum, pituitary gland, brain stem and spinal cord were ignored, while the measurements of the sphenoidal angle by Welcker in 1863, and those by Landzert in 1866, who had added that of a baboon, quantified the orientations of the clivus. They showed the angular discontinuity between humans and great apes, as well as between great apes and a small monkey like the baboon.

Georges Cuvier had nevertheless clearly highlighted the primary place of the nervous system in the classification, as well as the reasoning which consisted of starting from the general and moving to the particular, going from the interior (the axial endoskeleton in this case) to the exoskeleton, which offered a greater plasticity adaptive to the conditions of existence. The orientation of the cerebral trunk should have been the guiding principle of the nomenclature.

This being said, Paul Broca insisted on the structure that needed to be considered in the first approach in order to avoid errors induced by analogies:

> The form, considered in itself, would often yield the most misleading results. It would, for example, rank whales among fish. But the form subordinated to the structure, and applied, as Agassiz does, to the determination of groups that have a common structure, is the best characteristic of natural groups. (Broca, "L'ordre des primates, parallèle anatomique de l'homme et des singes", 1869)

He emphasized joints such as the wrist, with sustained attention to the fact that the human hand is not an organ of locomotion. It was already well-established that the skeleton and muscles of the hind feet of a great ape are not those of a hand, invalidating the expression "quadrumane". Great primates do have two feet and two hands; these serve as a support on the ground, not on the palm, but on the backs of the fingers thanks to their fairly long arms, while the feet have a function of supporting the body by resting on the plantar surface. The fingers resembled those of hands because of the grasping function when moving between the branches of trees.

He therefore distinguished:

> – the Hominian family. Vertical attitude; bipedal walking. Genera: *Homo*, Human.
>
> – the Anthropoid family. Synonymy: Anthropomorphs. Oblique attitude, close to vertical. Imperfect bipeds, usually taking an auxiliary support point on the dorsal side of the fingers when walking, not on the palms of the hands. Torsion of the humerus close to two right angles as in humans. Tail point. Organic structure extremely similar to that of a human being. Inhabit tropical Africa and the large islands of the Indian Archipelago. Genera: *Gorilla* (gorilla), *Troglodytes* (chimpanzee), *Satyrus* (orang), *Hylobates* (gibbon).
>
> – the Pithecian family. Synonymy: monkeys of the old continent, catarrhinian monkeys. Attitude closer to the horizontal than to the vertical, or completely horizontal. Quadruped walk, in which the front limb rests on the palm of the hand. Open nostrils below the nose (catarrhinian) (…). Inhabit the warm regions of the old continent, and Malaysia. Genera: *Semnopithecus* (*semnopithecus*), *Colobus* (colobus), *Cercopithecus* (cercopithecus or guenon), *Macacus* (macaque), *Inuus* (magot), *Cynocephalus* (dog-head or baboon).
>
> – the Cebian family. Synonymy: monkeys of the new continent or America, platyrrhinian monkeys. Attitude and walking as in the

> previous families. Open nostrils on the sides of the nose, which is flattened (platyrrhinian) (…). Inhabit the new continent. Genera: *Myceles* (hower monkey), *Ateles* (large spider-monkey), *Eriodes* (eriod), *Lagotrix* (lagotriche), *Cebus* (sajou), *Callitrix* (tamarin), *Nyctipithecus, Pithecia* (saki), *Saimiri*, *Hapale* (marmoset).
>
> – the Lemurian family. Synonymy: strepsirrhiniens, makis, false monkeys. Quadruped attitude and walk. Variable dental formula. Terminal and sinuous nostrils (strepsirrhinian) (…). Almost all live in Madagascar. Some of them live in East Asia or the Indian archipelago. Genera: *Lemur* (maki), *Stenops* (*Loris*), *Lichanotus* (*Indri*), *Tarsius*, *Galago* or *Otolichnus*, *Galeopithecus* and *Semnocebus* (Avahi). (ibid., author's translation)

Anthropomorphs – or anthropoids – are therefore closer to humans than monkeys in the degree of straightening. By integrating the brain and the convolutions, *Broca noted the asymmetry of the two cerebral lobes visible in chimpanzees and orangutans, so it was not specific to humans*. He held the coherence of its synthesis up to human intelligence:

> Thus, a slight organic improvement in itself can lead to various, numerous and profound functional consequences. This lack of proportion between the anatomical and the physiological change manifests itself in the development of the individual at the moment when an organ completes its evolution, and in the zoological series at the moment when an organ reaches, not the absolute perfection that no one can define, but a relative perfection, in relation to a determined function: This is why, in the parallel of man and anthropoids, the comparison of organs shows only slight differences, while the comparison of functions reveals much greater ones. And that is why I draw between these first two families of the order of primates a deeper demarcation than between the following families. (ibid., author's translation)

And to conclude:

> Man will be able to deploy and use his forces everywhere. He will no longer be confined to the forest, he will be able to roam the savannah, cross the steppes, inhabit the plain or the mountains, as he chooses, and become the conqueror of the entire planet. His hand, detached from the ground, will be a marvelous working instrument, an active instrument with which he will be able to create passive instruments, make and handle tools, offensive and defensive weapons. Able to run

> everywhere, he will be able to pursue and reach a living prey and add animal food to his vegetable diet. (ibid., author's translation)

Paul Broca thus drew up a picture of an argued classification of primates. Since then, the tarsier had caused controversy because of the suppression of the prosimia–simia distinction, imposed by the dogma of gradualism, or the missing link, which ignored discontinuities due to emergences and threshold effects, while they are characteristic of evolutionary processes. Today, the tarsier is brought closer to the monkeys (or Haplorrhines) through external criteria that hide the structure: contrary to appearances, the face is long like a lemur and not short as a monkey, but the exuberance of the orbits hides it, and the forwarded position of the occipital hole (*foramen magnum*) has nothing in common with the monkeys because there is no sphenoidal angle, unlike all monkeys at birth (details are given in Chapter 5 of Volume 2).

The grouping of Aristotle's anthropoids (macaques, baboons, etc.) no longer has anything to do with contemporary classification; it now corresponds to the family of the Pithecians. These are the Cercopithecids, with a morphology of molars recognizable among all of them.

10.2.5. *The transmission of acquired characteristics and Charles Darwin's gemmules*

In 1870, Charles Darwin presented his candidacy to the Anatomy and Zoology section of the *Académie des sciences* of the *Institut de France*, which was not followed up (Blanckaert 2015). In 1871, Charles Darwin wrote his first essay on the animal:

> The conclusion that man is the co-descendant with other species of some ancient, lower, and extinct form, is not in any degree new. Lamarck long ago came to this conclusion (…) as the progenitors of man became more and more erect (…) owing to a change in its manner of procuring subsistence or in the conditions of its native country, to live somewhat less in trees and more on the ground, its manner of progression would have been modified: and in this case it would have had to become either more strictly quadrupedal or bipedal. (Darwin, *The Descent of Man, and Selection in Relation to Sex*, 1871)

The British naturalist did not invent the theory of simian origins. He adhered to the Lamarckian theory and repeated it point by point: 1) the body of the monkey that transforms is the adult, 2) several bodies, male and female, would have straightened up because of the climate; the savannah was expanding, the density of trees was regressing, monkeys were forced to move longer standing between the trees (whereas quadrupedalism is more stable). The physical cause of straightening would thus be locomotion of the lower appendicular skeleton. Finally, 3) the anatomical modifications of the adult body would be transmitted to the seeds and preserved over several generations if the living conditions remained favorable, otherwise they would disappear.

Like Lamarck, Charles Darwin was not concerned with the internal organization of the base of the skull, and he had no idea of the physical impossibility for the muscles that attach the head to the spine to straighten the sphenoid which is at the center of the endocranium, covered with the meninges of the nervous system, without any contact with the muscles of the neck. The straightening can only come from the inside of the endoskeleton, in relation to the trajectories of brain development. However, not a single anatomist evoked this simple idea, even though skulls of human fetuses and very young orangutans had already been sketched after the extraction of the brain and described with the synchondroses of the base.

Isidore Geoffroy Saint-Hilaire died prematurely in 1861. Henri Milne-Edwards (1800–1885) succeeded him to the Zoology Chair of the Museum. In 1867, he expressed his thoughts on Charles Darwin's book by returning to the Lamarcko-Darwinian scientific impasse:

> Mr. Darwin's hypothesis does not seem to me likely to remove any of the difficulties relating to the origin of species, and in science it always seems to me annoying to mask our ignorance by alleged explanations which, in reality, explain nothing (…). As a result of a common defect among inventors, [Darwin] has exaggerated the power of the agent whose intervention he invokes and, in my opinion, something more is needed to explain the transformations of which paleontology offers us the picture (…). (Milne-Edwards, *Rapport sur les progrès récents des sciences zoologiques en France*, 1867).
>
> Darwin found his best defender in the person of Armand de Quatrefages, who had ensured his high reputation as a naturalist oceanographer before becoming, in 1855, the Museum's professor of anthropology. With the same pugnacity, he challenged in one stroke the suspicion of "amateurism" weighing on Darwin by recalling his contributions as a geologist and botanist and his talents as an observer. A skillful dialectician, Quatrefages was careful not to praise *The Origin of Species*. (Blanckaert, "Pour une théorie évolutive humaine", 2015, author's translation)

Charles Darwin probably finally admitted in 1868 the reason why naturalists of the scientific stature of de Quatrefages, Broca and Lartet remained indifferent. For a mind of the multidisciplinary stature of a Karl Kielmeyer, a transformation process must be scientifically transcribable by mathematicians and physicists. Charles Darwin was a long way from this. Not only did he not describe any physical process, but biologists had also understood well since Georges Buffon that heredity corresponded to very particular organic molecules; the notions of complex organization were explicit. It was the level of dynamic organization of hereditary complexity, from the content of gametes to the adult organism, that remained misunderstood. As long as this level of deployment of a complex dynamic from egg to adult was not conceptualized, the explanation of anatomical change, which was the process in the strict sense of transformation, was conceptually inaccessible. However, this was what was missing in Lamarck's and Darwin's reasoning: the discourse said nothing about the processes of hereditary transmission of acquired traits. The mechanisms sought were natural, but the word "selection" did not correspond to the physical reality that was supposed to replace metaphysics. Where were the forces with their amount of movement that would change the orientation of the bone trabeculae, the muscle insertions, the shape of the heart and arteries, the morphological and histological coherence of the joints, the straightening of the face, the lowering of the cerebellum? Where was the corpuscular, atomic or molecular reality, which would have abstraction capacities equivalent to metaphysical intelligence, that could transmit all these macroscopic systemic transformations to the cells of reproduction, on the microscopic scale of their organic molecules?

Darwin then imagined the existence of granules that he called "gemmules" (Darwin 1868), which would be present in all the cells of the body, male and female; they would be endowed with the properties of impregnation of the changes of each organ or tissue. Then they would converge together toward the ovaries and all the oocytes, thanks to blood circulation in vertebrates, in order to reconstruct there together the new physico-chemical combinations of the germ plasm (the gametes). However, this reasoning fails because organogenesis was not the deployment of an adult in miniature, as postulated by preformationists. This vision, which he called the "theory of pangenesis", implied as a minimum the same structural modifications between males and females for all organs; the statistics of sexual selection at least should have been taken into account. The founder of eugenics, his nephew Francis Galton (1822–1911), tested the hypothesis with experiments of blood transfusion to rabbits and invalidated it. He also conceived the existence of "gemmules" corresponding to the "hereditary memory of the species", but he distinguished himself from Charles Darwin by placing them first in the gametes, before they found their way into the cells of the organism, and he published an essay on the question, "Theory of heredity", in 1875 (Galton 1875). On November 7 1875 (letter 271), Darwin notified him that, every year, there was more evidence to confirm the Lamarckian theory he had made his own. Even though he did not explain how gemmules act in gametes, he was convinced of the veracity of his model and doubted that any one would ever follow Galton:

> If this implies that many parts are not modified by use and disuse during the life of the individual, I differ widely from you, as every year I come to attribute more and more to such agency (…). If I ask myself how you derive, and where you place the innumerable gemmules contained within the spermatozoa formed by a male animal during its whole life, I cannot answer myself – Unless you can make several parts clearer I believe (though I hope I am altogether wrong) that very few will endeavour or succeed in, fathoming your meaning. (Darwin, "Correspondence with Francis Galton", 1875)

Francis Galton tried to explain to Darwin that the absence of an organ did not come from lack of use, but from the repression of the gemmules in the gametes. However, Darwin did not visualize how these gemmules could have appeared in the gametes; they were necessarily deposited there from adult cells impregnated by the influences of the environment. It is not difficult to understand that, for Darwin, it was a question of coherence: the adequacy between anatomy, organs, functions and environment could not be a matter of chance and, when a new embryogenesis emerged, the "memory" could not "know in advance" to which constraints the postnatal stages would be exposed. It did not yet have the imprint of the environment.

The invention of gemmules was obsolete even before it was enunciated because, in 1865, the Czech monk Gregor Mendel (1822–1884) published his results on the study of the hybridization of peas, highlighting a law of distribution of the variation of the same trait resulting from the crossing of two parents (smooth and wrinkled peas) (Mendel 1866). Characteristics were not transmitted according to the vagaries of the environment, but according to rules specific to the cells of reproduction.

10.2.6. *The faults of Charles Darwin against Armand de Quatrefages*

In 1872, Charles Darwin reapplied to the *Académie des sciences*, but without success. Indiscretions revealed the content of opinions on the scientific deficiencies of his intellectual construction:

> Adolphe Brongniart, professor of botany at the Museum, degraded his system to the rank of a "fairy tale". During the same session, and without further amenity, the histologist Charles Robin is said to have declared "that Darwin should be appreciated on the basis of the demonstrable facts that he has introduced into science, and that according to this measure he should not even appear on the list (…)".

> Robin felt betrayed by this account. In a subsequent letter to the Review, he corrected and attenuated the form but kept the substance of the statement. In this concert of hatred and stupidity, the Museum could not remain in the background. (Blanckaert, "Pour une théorie évolutive humaine", 2015, author's translation)

However, the absence of demonstration by a histologist was well founded. If Armand de Quatrefages remained in his position, it was also because paleontological and anatomical facts required credible explanations, and nothing in Charles Darwin's remarks referred to the elaboration of a scientific theory, namely, the possibility of a mathematical transcription of the processes from adult to egg. In the second edition of 1872, Charles Darwin wrote:

> I now admit, after reading the remarks (…) made by Professor Broca, that in the earlier editions of my "Origin of Species" I perhaps attributed too much to the action of natural selection or the survival of the fittest (…). I did not formerly consider sufficiently the existence of structures, which, as far as we can at present judge, are neither beneficial nor injurious; and this I believe to be one of the greatest oversights as yet detected in my work (…) hence if I have erred in giving to natural selection great power, which I am very far from admitting, or in having exaggerated its power, which is in itself probable, I have at least, as I hope, done good service in aiding to overthrow the dogma of separate creations. (Darwin, *The Descent of Man, and Selection in Relation to Sex*, 1872)

Charles Darwin ended up focusing his attention on embryology, more as a strong argument in favor of evolution, as Geoffroy Saint-Hilaire had long argued, than as the first evolutionary problem already solved by his theory on gemmules.

> The discovery is that the larvæ of Ascidians[1] are related to the Vertebrata, in their manner of development, in the relative position of the nervous system, and in possessing a structure closely like the *chorda dorsalis* of vertebrate animals. It thus appears, if we may rely on embryology, which has always proved the safest guide in classification, that we have at last gained a clue to the source whence the Vertebrata have been derived (…). We shall be able to reconstruct partially during successive periods the structure of our early

1 "The ascidians belong to the branch of the Prochordata, to the sub-branch of the Urochords or Tuniciers. It was the Russian embryologist Kowalevsky who demonstrated, thanks to embryological studies on simple Ascidians from 1868 to 1872, the close relationship between ascidians, Amphioxus and Vertebrates" (Chebbi 2010).

> progenitors. This can be effected by means of the rudiments which Man still retains, by the characters which occasionally make their appearance in him through reversion, and by the aid of the principles of morphology and embryology (...). Each organism will still retain the general type of structure of the progenitor from which it was aboriginally derived (...). If we turn to the geological evidence, that organisation on the whole has advanced throughout the world by slow and interrupted steps. (Darwin, *The Descent of Man*, 1872)

In a letter dated July 9 1873, despite the failure of his candidacy, he thanked Armand de Quatrefages for having intervened in his favor:

> My dear Sir, I am much obliged for your very kind letter and exertions in my favour. I had thought that the publication of my last book would have destroyed all your sympathy with me; but though I estimated very highly your great liberality of mind it seems that I underrated it. I am gratified to hear that M. Lucaze du Thiers, will vote for me, for I have long honoured his name. I cannot help regretting that you should expend your valuable time in trying to obtain for me the honour of election; for I fear, judging from the last time that all your labour will be in vain. Whatever the result may be, I shall always retain the most lively recollection of your sympathy and kindness; and this will quite console me for my rejection. P.S. With respect to the great stress which you lay on man walking on two legs, whilst the Quadrumana go on all fours, permit me to remind you that no one much values the great difference in the mode of locomotion and consequently in structure between seals and the terrestrial Carnivora, or between the almost biped Kangaroos and other marsupials. (Darwin, "Correspondence with de Quatrefages", 1873–1879)

It was at the request of Charles Darwin that the holder of the Chair of Anthropology of the Museum supported his candidacy for his skills in zoology and botany. The postscript reveals how Darwin was unaware of the internal structural differences between the neurocranium of *H. sapiens* and that of a kangaroo. Under the influence of the Lamarckian school, reducing the origin of the human anatomy to locomotion, the British naturalist masterfully missed the evolutionary problem. The neurocranium of the kangaroo is as flat as that of a carnivore or a lemur. Armand de Quatrefages had solid reasons for not giving him the change. Charles Darwin obviously had a very high opinion of his model, as revealed by the expression "let me remind you", whose professorial tone reveals a certain sufficiency with regard to the skills of Armand de Quatrefages, physician, anthropologist and geophysicist.

De Quatrefages granted the organization of humans the level of the order of quadrumanes. This conclusion was consistent with the mode of reasoning used in the Linnaean nomenclature. The director of the Anthropology Laboratory was free, at the Museum, at the Anthropological Society of Paris and at the *Académie des sciences* not to consider *On the Origin of Species* as the discovery of the century; Charles Darwin was not the Newton of the 19th century. He never resolved the most complex of questions, namely, the relations between the evolutionary processes that governed the reorganization of embryogenesis and the emergence of the anatomy and physiology of *H. sapiens*, and finally, structured the capacities of this organism to consciously question its links with the natural history of the universe.

Charles Darwin did not express the wish to consult the collections of paleontology and comparative anatomy preserved in the *Jardin des plantes*, to observe the rare original fossils of monkeys, with the mandibles of *Pliopithecus* and *Dryopithecus*, and compare them with the mandibles of current species, the chimpanzee, gorilla and orangutan and *Hylobatidae* (gibbon family) – and, why not, compare them placed back in their skull base. Nothing in *The Descent of Man* resembles a scientific theory based on the metric study of the growth trajectories of the base of the skull, whereas, according to Darwin, it should have been central. The historical reality was iconoclastic, Darwin had no knowledge of the structural differences between the endocranial base of a human being and that of a quadruped, whether a kangaroo, which also relies on its two forelimbs, a seal or a lemur, which all have a flat cranial base. The scientific posture of de Quatrefages escaped him because the origin of humans was not a question of locomotion, of adaptation of the appendicular skeleton to the need to straighten up. Darwin was not elected until 1878, four years before his death.

Broca agreed with Huxley, demonstrating the greater proximity between chimpanzee and humans than between chimpanzee and gibbon. However, he never interpreted the differences in structures related to verticality as the result of locomotor adaptation of the appendicular skeleton. In this, he proved Armand de Quatrefages right. It took much more than words like "natural selection of the most adapted by the most adapted" to account for a transformation of bone structure. It was necessary to have the height of vision of an anthropologist like de Quatrefages, to understand that a mechanical locomotor cause was powerless to transform the internal and external organization of the skeleton. No knowledge in biophysics could explain all the structural changes until their transmission by gametes. Moreover, Darwin did not return to the Lamarckian dogma that he made his own, and therefore to his theory of pangenesis, despite the experiments that invalidated it.

With his gemmules, Darwin proposed a kind of "self-memorizing morphodynamic" principle but without being in a position to give a physical equivalence on which a physicist, a chemist or a mechanic could have relied. In spite of the notable absence of explanation of the processes supposed to change the adult forms and then transmit them to the ovaries in females, the expression "natural selection" would pass into everyday language, qualified as an "engine", whereas it was in fact a metaphor. This term does not describe any work: there was no energy source, no force, no amount of movement and no dynamics. The word "engine" hides the absence of a genuine scientific discovery regarding the processes of morphogenetic transformation since fertilization and the impossibility of writing the equation of the trajectories of emerging forms to establish their rationality.

11

Transformist Paleontology Inaugurates the 20th Century

11.1. The rebirth

The National Museum of Natural History created the first Chair of Paleontology in 1853 for Alcide d'Orbigny (1802–1857), who throughout his career encountered a lack of recognition of the discipline, which was deemed unnecessary. Following his death, it was left vacant for a good 10 years until Edouard Lartet took it over in 1869 for only two years, as he in turn died in 1871. The Museum did not have a Chair of Embryogeny; Isidore Geoffroy died prematurely in 1861, Étienne Serres died in 1868 and paleontology, frozen in a nomenclature without an evolutionary vision, did not arouse passions. Alfred d'Orbigny's chair was then entrusted in 1872 to his brother-in-law Albert Gaudry (1827–1908).

11.1.1. *Albert Gaudry (1827–1908), a palace in the* Jardin des plantes *for paleontology and comparative anatomy*

Albert Gaudry, who joined the Chair of Paleontology as a "naturalist's assistant" in 1853, knew what was at stake. He invested in a vast financial program to extricate the Museum from Cuvier's dogmatism and to develop the exceptional paleontological potential of the national collections. His project was the construction of a monumental gallery of paleontology and comparative anatomy in the *Jardin des plantes*, going up the Seine to the former *Jardin* royal castle. Gaudry succeeded in his challenge and unleashed the transformism of Lamarck and Geoffroy Saint-Hilaire in the heart of Paris, in a masterful demonstration of the richness of the collections and their educational and didactic value. The museum's design was thought out in such a way that the comparative anatomy of current and fossilized species of plants, invertebrates and vertebrates, spoke for itself to the general public, following the great geochronology fresco. Now citizens could come and see for themselves the natural history of the evolution of life, plants and animals. The spectacle of skeletons walking toward the visitor who entered this great exhibition of origins officialized the recognition of evolutionary paleontology in the heart of the French capital. The inauguration took place in 1898.

The evolutionary continuum between fossil species has therefore been taught in France, thanks to Albert Gaudry, since 1871. A convinced evolutionist, he was nonetheless a man of faith and as such he expressed reservations about the concept of natural selection if it were to be reduced to the "law of the strongest". He enunciated the "law of love" in 1902 (Gaudant 2008). The evolution of organizational plans related to complex physico-chemical processes that had disappeared and were irreversible. For the paleontologists who conceived them, these processes could only be approached by understanding the ontogeny of current species that had preserved their foundations, at the level of structures elevated to the rank of order, class or branch, shared with fossils. The great unknown was the metabolism of sex cells, where the "memory" of the reconstruction of organizational plans was condensed.

11.1.2. *Paul Gervais (1816–1879) at the Chair of Anatomy and the first bipedal fossil monkey*

Étienne Serres's successor to the Chair of Comparative Anatomy was not an embryologist but a zoologist. Paul Gervais (1816–1879) was a former student of the Museum; of modest origin, he was trained on the job. He was already 25 years old when he attended classes at the *Jardin de Plantes*. He financed his studies by starting as an assistant to the Chair of Comparative Anatomy, then as a tenured assistant in 1841. He obtained his doctorate in 1844. Back at the Museum after a career in Montpellier, he organized paleontological prospecting missions abroad. In 1870, these were a success for paleoprimatology, so poor in specimens. Paul Gervais discovered in Tuscany, the country of Leonardo da Vinci, an incomplete skeleton of an unknown genus of monkey that he named *Oreopithecus bambolii.* The 7 to 8 million year old skeleton was collected in lignites in the commune of Montebamboli. The fossil aroused interest because of its preservation. The great length of the arms recalled that of a modern gibbon, and the posterior appendicular skeleton, which included the pelvis and the two femurs, showed bipedal locomotion as in gibbons. Its crushed skull had preserved the mandible; it had nothing comparable to humans. We will see that it was indeed similar to this genus and resembled the *Pliopithecus* of the south of France. The fossil preserved in the Museum of Paleontology in Florence has been the subject of a recent scan study. The authors concluded:

> We correct several crucial interpretations relating to the anterior inferior iliac spine and lumbar vertebrae structure and identifications (…). O. bambolii lacks *the complete reorganization for torso stiffness seen in extant great apes* (i.e., living members of the Hominidae), and is more similar to large hylobatids in certain aspects of torso form. (Hammond *et al.*, "Insight into the lower torso in late Miocene hominoid Oreopithecus bambolii", 2020)

This bipedalism clearly showed that it did not modify anything of the cranial structures, notably the mandibular structures. The differences in form in the three plans bringing together all the monkeys, whether American (Platyrrhinian) or Old World (Catarhinian), they were placed at a very high level of classification, well before the great apes, that is to say at a very old axial organization plan, at least 39–40 million years old. The method of comparing the current morphogenesis of primates and each stage preserved in the fossil state reinforced Étienne Serres's logical reasoning. In order to understand the differences between the axial structure common to the current great apes and fossils, which included the mandible, and that of small apes, it was necessary to look at the classification that brought them together. What fixists did not understand was how the embryonic morphogenetic plan "monkey", common to current and fossilized species, could be modified to that of current and fossilized great apes.

The interest in this fossil found an echo especially in Germany, where Goethe's evolutionism eventually found an emulator in evolutionary embryology with Ernst Haeckel. He published a reflection on the relations between embryogenesis and phylogenesis in 1866 after visiting Charles Darwin, Thomas Huxley and Charles Lyell in London to discuss the subject.

11.1.3. *Haeckel (1834–1919), on the way to formalizing processes*

Ernst Haeckel was the great promoter of biological evolution in Germany, demonstrating the close relationship between the embryogenesis of current classes and their phylogenesis. He studied medicine and natural sciences in Würzburg (Bavaria, Germany), where he worked for some time as an assistant to Rudolf Virchow. He discovered the work of Meckel and von Baer and undertook his doctoral thesis in 1857 and 1858 on the anatomy of the crab. In 1862, he was appointed extraordinary (i.e. temporary) Professor at the Chair of Comparative Anatomy at the Institute of Zoology of the University of Jena. In 1866, he published his own conception of evolution without having discovered its processes. His book is entitled *Generelle Morphologie der Organismen. Allgemeine Grundzüge der Organischen Formen-Wissenschaft, Mechanisch Begründet durch die von Charles Darwin Reformirte Descendenz-Theorie* (General Morphology of Organisms:

General Principles of the Science of Organic Forms, Mechanically Founded by Charles Darwin's Reformed Descendence Theory).

It is a sum of reflections and syntheses with new concepts published in two volumes, marking a turning point in the formulation of the theory of evolution and the origins of humans. His reflection was very accurate in its innovative lines since he understood, as he wrote, that force and matter were identical.

The whole of his work was driven by the need to obtain freedom from biblical dogma in order to understand the natural links between the evolution of forces from at least the origins of life through the physiology of the human and to human consciousness. He made it clear how much the awareness of the place of human embryogenesis, and in particular of neural embryogenesis, in evolution was the equivalent of the Copernican revolution. Humans had not experienced a greater shock since Copernicus:

> The development of man is therefore the result of the same laws as the evolution of any other body of nature. By establishing this monistic notion scientifically and definitively, our era has made immense progress in the unitary conception of the world. Only one other intellectual conquest can rival this one, and that is the one Copernicus made four centuries ago by ruining Ptolemy's world system. By demonstrating that the earth was not the center of the world, but a grain of dust in the universe, Copernicus destroyed the ancient geocentric concept and created a new system of the world, to which Newton gave a mathematical basis through his theory of gravitation. Similarly, at the beginning of our century, *Jean Lamarck ruined, through his theory of descent, the anthropocentric conception of the world, then dominant and according to which man was the center, the goal of creation.* (Haeckel, *Natürliche Schöpfungsgeschichte*, 1875, author's translation)

As a thoughtful and well-trained scientist, he understood the problem of force and form, the dilemma between the model (or archetype), and the real world which was merely displacement from atoms, but ordered displacement, organized at all scales of organization of crystals and living beings. To do this, he developed a geometrical study of shapes in three dimensions, which he called stereometry, tracing triangles or lozenges (see Taff II in Haeckel (1866), available online).

During a trip to the Canary Islands in 1866 and 1867 via England, Haeckel visited Charles Darwin, Thomas Huxley and Charles Lyell. He returned to Germany impressed by Darwin, who convinced him to make the importance of evolution known in order to become aware of the place of humans in the animal kingdom. It was important, however, to place this enthusiasm in the geopolitical context of a Europe undergoing colonial expansion and industrial revolution, driven by a racial and hierarchical vision of human biological diversity, cultures and civilizations.

In France, Isidore Geoffroy had died two years previously, Étienne Serres was still alive and Victor Coste still held the Chair of Comparative Embryogeny at the *Collège de France*. Haeckel knew how to integrate his works and illustrations and recognized the primacy of Lamarck and then Geoffroy Saint-Hilaire:

> The theory of descent, considered as a great branch of universal evolutionary theory, was already clearly formulated by Lamarck, and pushed by him to its most important consequences, so much so that he must be considered as the true founder. *It is therefore necessary to call Darwinism not the theory of descent, but the theory of selection.* This last theory is of such importance that it cannot be estimated too highly. (Haeckel, *Histoire de la création des êtres organisés d'après les lois naturelles*, 1874, author's translation)

Following his meeting with Darwin, Haeckel was convinced of the relevance of the evolution of fossilized species but noted the absence of embryogenesis in his reflection. Moreover, Darwin had not yet written an essay on the origins of humans.

11.1.4. *Haeckel, a new hope*

For the French teaching staff, Darwin's "reformed descent theory" had not reformed anything. Lectures at the University of Jena were completed in 1874 and translated into French in 1875. The author of the preface, Charles Martins (1808–1889), let himself be carried away by Haeckel's enthusiasm, letting readers believe in the reality of an unprecedented discovery explaining the appearance of new species. This "discovery" was portrayed as a victim of the contempt of the French intelligentsia, perceived as incapable of recognizing a genius:

> For lack of sufficient material borrowed mainly from paleontology, for lack of numerous proofs drawn from comparative anatomy of morphology and embryology, Lamarck and Geoffroy Saint-Hilaire had not convinced their contemporaries; Darwin was happier (…). Most of the numerous writings, which during this period were published for or against Darwinism, were the work of people who lacked sufficient biological and especially zoological instruction. (Martins, foreword to Haeckel, *Histoire de la création des êtres organisés d'après les lois naturelles*, 1875, author's translation)

In reality, Darwin did not proceed with any scientific theory, as Haeckel claimed it existed, involving the discovery of physical laws. French scientists were not fooled. Darwin and Huxley were no more qualified in comparative paleontology, comparative embryogeny and anthropology than Henri de Blainville, Pierre Flourens, Étienne Serres, Victor Coste or Armand de Quatrefages. Their skills made it possible for them to reject the "theory" of the hereditary transmission of acquired traits as the solution to the evolutionary problem of organizational plans that Darwin had made his own. However, the most hostile to Lamarckian transformism, like de Quatrefages, did not hinder the recognition of his naturalist skills. On the contrary, the French anthropologist had allied himself with Félix Lacaze-Duthiers (1821–1901) to support his candidacy for the *Académie des sciences*. Félix Lacaze-Duthiers was a doctor at the Faculty of Paris, after having been a student of Henri de Blainville, then the successor to the chair of Isidore Geoffroy Saint-Hilaire and a member of the *Académie des sciences*. Likewise, Charles Martins seemed to have forgotten that Albert Gaudry, holder of the Chair of Paleontology at the Museum and Director of the Museum's paleontology laboratory since 1871, was a convinced evolutionist.

In Paris, evolution was thus recognized by Albert Gaudry and by Paul Broca, whose reputation streatched across Europe. Evolution was accepted in the sense of a filiation between fossil and current forms, through natural processes, but morphogenetic processes were misunderstood. Evolution was therefore officially accepted in France, at the National Museum of Natural History in Paris, and taught as a necessary doctrine based on the observation of adaptations to living environments, without denying the increasing complexity of organizational plans, for which the organ most directly concerned was the central nervous system. It was therefore up to the young Haeckel to write a new page in the theory of evolution by discerning what differentiates the organogenesis of the species of each phylum and of each class and comparing fossils following the geochronological scale. The intellectual context that prevailed in Germany was guided by the thought of Leibniz and Goethe, who made the break with scholasticism. This doctrine was monism, which removed ontology and metaphysics from reflection:

> We must definitely accept the conception of the universe as mechanical or causal. We can also call this the monistic or unitary way of seeing as opposed to the dualistic view implicitly contained in any teleological explanation of the world. (Haeckel, *Histoire de la création des êtres organisés d'après les lois naturelles*, 1874, author's translation)
>
> The common fundamental ideas of cosmic unity, of the inseparable solidarity of force and substance, of spirit and matter, or, as it can also be said, of God and the world. No one has given this conception a

> more poetic expression than the greatest of our poets and thinkers, Goethe, in his Faust and in his wonderful poem God and the World. (Haeckel, *Le monisme, lien entre la religion et la science*, 1897, author's translation)

11.1.5. *Phylogenesis and embryogenesis, a reversed logic*

Back at the University of Jena, Haeckel was less than 40 years old when he tried to convince his contemporaries of the relevance of evolution. The vast majority had seen no visual representation of the human embryo, nor had any idea of their own very close anatomical proximity at four embryonic weeks to the dog or to a lamprey in the even earlier stages. This was, in his eyes, the great lesson of humility for theologians, forced to see themselves as closer to the lamprey from the very beginning of their existence than to an Architect whose genius went beyond the dimension of the cosmos. His intention was to show not only the animality of humans, but also their physiological community with all organisms and without any supernatural exception.

His lectures took up what Lamarck had already published in 1794 with his "physico-chemical logic":

> The great importance of the history of evolution for the scientific intelligence of the world of animals and plants has been so generally recognized for some decades that, without it, it is impossible to take a somewhat assured step in organic morphology, *in the science of forms*. However, the expression "history of evolution" has almost never been understood to mean a fragment of this science, that is to say the evolution of organized individuals, which is usually called *embryology,* and which would be better designated by the more accurate and more comprehensive expression of *ontogeny*. However, apart from the other great biological problems, the doctrine of descent provides us with answers that are not only satisfactory, but also have the merit of attributing only to natural mechanical causes, to physico-chemical forces, phenomena that, for a long time, it was customary to attribute to supernatural creative forces. (Haeckel, *Generelle Morphologie der Organismen*, 1866, author's translation)

Haeckel innovated by bringing together ontogenesis and phylogenesis in his famous formula "ontogenesis sums up phylogenesis". His retrospective consisted of placing human embryogenesis in this great phylogenesis of organizational plans.

> But, apart from this science (ontogeny), there is also a history of the evolution of species, classes and organic families, and this history is related to the first one in extremely important ways. The materials of this history are provided to us by paleontology. This science teaches us that, during the multiple periods of the Earth's evolution, each group of animals and plants has successively passed through a whole morphological series of very diverse classes and species. The group of vertebrates, for example, has passed through the class of fish, amphibians, reptiles, birds and mammals, and each of these classes has also passed through a series of varied species. Now this history of the paleontological evolution of organisms, *which can be called the history of organic families or phylogeny,* relates in the most important and remarkable way with the other branch of the history of organic evolution, the one that deals with the individual, the ontogeny. (Haeckel, *Histoire de la création des êtres organisés d'après les lois naturelles*, 1874, author's translation)

From this point of view, the phylogenetic retreat to replace the human form was absolutely essential, the reality remained unchanged.

> As my "Anthropogeny" is the first attempt to expose ontogeny and human phylogeny in their etiological connection, I fear that my work leaves much to be desired. But, whatever it is, it will be enough, I hope, to convince any intelligent reader that, in order to become a true science, human embryology must go back to the causes of the phenomena. It is phylogeny alone that can elucidate ontogeny, because it reveals to us the true causes. (Haeckel, *Anthropogénie ou histoire de l'évolution humaine*, 1877, author's translation)

Haeckel published illustrations by Coste with the face of the human embryo in formation; it was not that of an adult prosimian, then a monkey, a great ape and an extinct anthropomorphic like the Neanderthal.

Whatever the species, the embryonic structure had always been embryonic, it had always preceded the more elaborate stages up to reproductive maturity.

Ernst Haeckel's reasoning was reversed, it was the evolution of embryogenesis that was the cause of phylogenesis.

11.2. Natural selection and the scale of human societies

This inversion can be explained by its attachment to natural selection acting on the adult stages. This was a strange paradox since, in Darwin's mind, natural selection only made sense on the condition that it attributed to the environment the power to shape the adult body. Only males and females whose gemmules would have transmitted the modifications to the spermatozoa and oocytes were reproducible, the unsuitable individuals eventually disappearing. In reality, Haeckel was guided by the ideal of a European society built on the notion of superior race. In 1905, he became an honorary member of the "Society for the Hygiene of Race" (*Gesellschaft für Rassenhygiene*), which was driven by a civilizational project. In 1909, he created the Monist League (*Deutscher Monistenbund*), presided over in 1910 by the winner of the Nobel Prize for Chemistry, Whilem Oswald (1853–1932). Monism considered space-time-matter-energy as a creative spiritual substance and cerebral matter was the highest elaboration of it in the evolutionary course of life.

What preoccupied Haeckel, as well as his eugenicist contemporaries, was more a concern with the conservation of the race, with seeking the success of its survival, namely, in its natural and cultural living conditions, than a real concern to understand the processes that transformed embryos of *Dryopithecus* into homininae, from sexual cell lines that went back to the depths of time. Haeckel's incoherence was pointed out.

The embryologist and morphologist Louis Vialetton (1859–1929), the first Professor of Histology at the Faculty of Medicine of the University of Montpellier (Vialetton 1924), reviewed the numerous criticisms by embryologists of Haeckel's recapitulation theory (Vialleton 1908). In short, it was a montage aimed at validating Darwinism as Haeckel defined it; as Vialleton described it, it was:

> Embryological interpretation based on an imperfect knowledge of things, and on the admission of a "fabulous heredity that confounds the imagination". (Vialleton, *Un problème de l'évolution*, 1908, author's translation)

The embryologist Vialleton was not against evolution but as a structuralist he noted the impotence of Lamarcko-Darwinism to account for the appearance of internal structures through slow processes constructed by the environment. He was part of a current that recognized the need for mechanisms capable of restructuring the unity of the organizational plan and its complexity, but these processes were still unexplained.

Since regulation of the genetic control of embryogenesis was not even conceptualized, Vialleton's position was legitimate. It was not anti-evolutionary, it was opposed to the transformist illusion, which was not a mechanism describing modifications of molecular coordinations and cell divisions since fertilization:

> It is perfectly possible that species of the same type, simple replicas of the latter, hardly modified in their superficial and accessory parts, are born one from the other by a differentiation resulting from the action of natural causes. However, it cannot be the same for types whose anatomy, being too different, do not lend themselves to such changes. It is generally agreed that the power of transformist factors stops at the formation of inferior types (species, genera, or at least certain families), that it is completely incapable of producing superior types (orders, classes, branches). The explanation for the genesis of species does not apply to that of types: the problem of the specification or adaptation of an existing type to diverse conditions of time, place or circumstances; the problem of organization, which concerns the very basis of the anatomical constitution, otherwise stable, otherwise rigid than the facts belonging to the domain of specification. This is what we wanted to mark by the very title of this book. (Vialleton, *L'origine des êtres vivants, l'illusion transformiste,* 1929, author's translation)

The complexity of the problems of generation and heredity exceeded contemporary knowledge and Charles Darwin was the first to recognize this, but with the deep conviction that the future would prove him right. His theory of pangenesis was challenged in 1883 by Auguste Weismann (1834–1914), a physician by training, teacher at and then director of the Zoological Institute of the University of Freiburg in Breisgau (southwestern Germany).

Weismann found that worker ants do not reproduce, so the specific characteristics that distinguish them, and that have been passed on for millions of years, could only come from the cells of reproduction. He then supposed that the fertilized oocytes divided into two categories of cell lines, one that transmitted the memory of the species, or germen, and the other that developed the organism, or soma. He replaced the theory of germ plasma with the theory of the "hereditary substance of extremely complex organization" (Weismann 1892, author's translation). It was contained in the nucleus of genital cells and was clearly separated from the soma.

At the end of the 19th century, when positivism was trying to impose itself as a new religion, naturalists were invited not to confuse science with reality. The clear-sightedness of Albert Gaudry, mastermind of the monumental gallery of paleontology and comparative anatomy, alerted his contemporaries to the need for vigilance and opposed to the law of selection the "law of love", which was transmitted from generation to generation and had nothing in common with the imperative to "select one another".

Paul Broca too had developed over the years and had already concluded in 1871:

> What astronomers, physicists and chemists did in the inorganic field, what organic physiologists did in biology, transformism strives to do in turn in natural history.
>
> To show that *the evolution of organic forms*, the appearance of species, their extension, their extinction, their succession and their distribution are ordinary phenomena, that is to say necessary and governed by laws which leave no room for a higher power, such is the aim, or at least such is the consequence of this hypothesis, whose boldness astonishes or enrages even many spirits attached to the most widespread beliefs, but which, by the same token, easily attracts to it spirits impatient to escape the yoke of dogmas (…).
>
> The nature of the relations of continuity that one establishes between the various terms of the organic series is certainly of the utmost importance; but what is even more important is this general notion that *the innumerable forms of organized beings are not distributed in nature in a capricious and confused way*, that they are subjected to a regular ordination, that they are consequently regulated by laws. It seems that such a notion could not spread throughout natural history without immediately dragging the spirits in search of these laws (…). What I would like to dream of for future humanity, others, no doubt, would like to grant to nascent humanity; *but science is not made to obey our tastes nor to flatter our pride, and it would be no more unfortunate to incline it before a theological system than to place it at the service of the philosophical doctrine which is in struggle with that system*. (Broca, "Sur le transformisme", 1870, author's translation)

References

Annoni, J.-M. and Barras, V. (1993). La découpe du corps humain et ses justifications dans l'antiquité. *Canadian Bulletin of Medical History*, 10(2), 185–227.

Arasse, D. (1997). *Léonard de Vinci*. Hazan, Paris.

Aristotle (1910a). *The History of Animals*, translated by D'Arcy Thompson. Clarendon Press, Oxford.

Aristotle (1910b). *On the Generation of Animals*, translated by Arthur Platt. Clarendon Press, Oxford.

Aveline, D.C., Williams, J.R., Elliott, E.R., Dutenhoffer, C., Kellogg, J.R., Kohel, J.M., Lay, N.E., Oudrhiri, K., Shotwell, R.F., Yu, N., Thompson, R.J. (2020). Observation of Bose–Einstein condensates in an Earth-orbiting research lab. *Nature*, 582, 193–197.

von Baer, K.E. (1827). *Ovi Mammalium et Hominis genesi, epistolam ad Academiam Imperialem Scientiarum Petropolitanam*. L. Vossius, Lipsiae, Leipzig.

von Baer, K.E. (1828–1837). *Entwickelungsgeschichte der Thiere : Beobachtung und Reflexion*. Bornträger, Königsberg.

Balan, B. (1979). *L'ordre et le temps. L'anatomie comparée et l'histoire des vivants au XIXème siècle*. Vrin, Paris.

Barthélémy-Saint-Hilaire, J. (1883). *Histoire des animaux d'Aristote*. Hachette, Paris.

de Bayle, P. (1697). *Dictionnaire historique et critique*, volume 4. Reinier Leers, Rotterdam.

Beaurepaire, M.C. (1866). *Georges Cuvier, Secrétaire-greffier de la commune du Bec-au-Cauchois*. Imprimerie impériale, Paris.

Belon, P. (1555). *L'Histoire de la nature des Oyseaux, avec leurs descriptions et naïfs portraicts retirez du naturel, escrite en sept livres par Pierre Belondu Mans avec privilège du Roy*. Gilles Corozet, Paris.

Berengario da Carpi, J. (1521). Commentaria cum amplissimus additionibus super anatomia Mundini. Hyeronimum de Benedictis, Bologna [online]. Available at: https://archive.org/details/ita-bnc-mag-00001056-001/page/n20/mode/2up.

Berengario da Carpi, J. (1522). *Anatomia Carpi : isagoge breves, perlucide ac uberime, in anatomiam humani corporis, a communi medicorum academia usitatam*. Bernardinum de Vitalibus Venetum, Venice.

Birembaut, A. (1957). Fontenelle et la géologie. *Revue d'histoire des sciences*, 10–4, 360–374.

Blanckaert, C. (2015). Pour une théorie évolutive humaine. Armand de Quatrefages, la formation des races et le darwinisme au Muséum national d'histoire naturelle. *Revue d'histoire des sciences humaines*, 27, 189–230.

Blumenbach, J.F. (1795). *De generis humani varietate nativa*. Vandenhoek and Ruprecht, Göttingen.

Blumenbach, J.F. (1798). *Über die natürlichen Verschiedenheiten im Menschengeschlechte*. Breitkopf and Härtel, Leipzig.

Bontius, J. (1658). *Historiae Naturalis & Medicae Indiae Orientalis*. Apud Ludovicum et Danielem Elzevirios, Amsterdam.

Borelli, G.A. (1680). De motu animalium. Opus posthumum [Online]. Available at: https://gallica.bnf.fr/ark:/12148/bpt6k512312/f413.item.

Bourdier, F. (1960). Trois siècles d'hypothèses sur l'origine et la transformation des êtres vivants (1550–1859). *Revue d'histoire des sciences et de leurs applications*, 1(13), 1–44.

Broca, P. (1858–1859). Mémoire sur l'hybridité et sur la distinction des espèces animales. *Journal de physiologie*, 1, 432–471 and 684–729, 2, 218–250 and 345–390.

Broca, P. (1859). Sur les races primitives contemporaines de l'époque dite du Diluvium. *Bull. Soc. d'Anthropologie de Paris*, 1, 70–76, 86–92.

Broca, P. (1859–1860). Mémoire sur les phénomènes d'hybridité dans le genre humain. *Journal de physiologie animale et humaine*, 2, 601–625, 3, 392–439.

Broca, P. (1865). Sur un moyen de mesurer l'angle sphénoïdal sans ouvrir le crâne. *Bull. Soc. d'Anthropologie de Paris*, 6, 564–572.

Broca, P. (1869). L'ordre des primates, parallèle anatomique de l'homme et des singes. *Bulletins et Mémoires de la Société d'Anthropologie de Paris*, 4, 228–401.

Broca, P. (1870a). Sur le transformisme. *Bulletins et Mémoires de la Société d'Anthropologie de Paris*, 5, 168–242.

Broca, P. (1870b). *Histoire des progrès des études anthropologiques depuis la fondation de la Société*. Hennuyer, Paris.

Buffon, G. (1749). *Histoire naturelle de l'Homme*. Imprimerie royale, Paris.

Buffon, G. (1769). *Histoire naturelle, générale et particulière*, volume 3. Imprimerie royale, Paris.

Buffon, G. (1785). *Histoire naturelle, générale et particulière*, volume 6. Imprimerie royale, Paris

Buffon, G. (1830). Expériences au sujet de la génération. In *Œuvres complètes, Histoire des Animaux*, volume 11. Pillot, Paris.

Buffon, G. (1838). *Œuvres complètes par Geoffroy Saint-Hilaire*, volume 2. Pillot, Paris.

de Burtin, F.X. (1784). *Oryctographie de Bruxelles, ou description des fossiles tant naturels qu'accidentels découverts jusqu'à ce jour dans les environs de cette ville*. Imprimerie Le Maire, Brussels.

Byl, S. (1997). Controverses antiques autour de la dissection et de la vivisection. *Revue belge de philologie et d'histoire*, 75, 1.

Camper, P. (1791). *Naturgeschichte des Orang-Utang und einiger andern Affenarten, des Africanischen Nashorns, und des Rennthiers Ins Deutsche uberstetzt, und mit den neuesten Beobachtungen des Verfassers herausgegeben, von JFM Herbell*. Johann Christian Daenzer, Dusseldorf.

Cardano, G. (1556). *De la subtilité et subtiles inventions, ensemble les causes occultes, et raisons d'icelles*, translated by Richard Le Blanc. Charles Langelier, Paris.

Carnot, S. (1824). Réflexions sur la puissance motrice du feu et sur les machines propres à développer cette puissance. *Annales de l'École Normale*, 2(1).

Cavaillé, J.P. (2001). Le prince des athées. In *L'enjeu Machiavel*, Cavaillé, J.P. (ed.). Presses Universitaires de France, Paris.

Cazes, H. (2019). Le corps des livres (I), une allégorie de l'anatomie féminine (Jacopo Berengario da Carpi, Commentarii… 1521). Blog of the Société bibliographique de France [online]. Available at: https://histoirelivre.hypotheses.org/4347.

Chebbi, N. (2010). Etude systématique, bio-écologique et chimique des acscidies de Tunisise. PhD thesis, National Agronomic Institute of Tunisia, Tunis.

Charbonnel, J.R. (1919). *La pensée italienne au XVIe siècle et le courant libertin*. Champion, Paris.

Coiter, V. (1572). *Externarum et internarum principalium humani corporis*. Theodorici Gerlatze, Nuremberg [Online]. Available at: https://archive.org/details/hin-wel-all-00000250-001/page/n88/mode/2up.

Condorcet, J.-A. (1790). *Éloge de M. le Comte de Buffon*. Buisson, Paris.

Cuvier, G. (1811–1828). *Biographie universelle ancienne et moderne, histoire par ordre alphabétique de la vie publique et privée de tous les hommes*. Michaud, Paris.

Cuvier, G. (1812). *Recherches sur les ossemens fossiles de quadrupèdes, ou l'on rétablit les caractères de plusieurs espèces d'animaux que les révolutions du globe paroissent avoir détruites. 1. Le discours préliminaire et la géographie minéralogique des environs de Paris*. De Terville, Paris.

Cuvier, G. (1830). *Discours sur les révolutions de la surface du globe, et sur les changemens qu'elles ont produits dans le règne animal*. Edmond d'Ocagne and G. Dufour, Paris and Amsterdam.

Cuvier, G. (1841–1845). *Histoire des sciences naturelles depuis leur origine jusqu'à nos jours, chez tous les peuples connus*. Fortin, Masson et Cie, Paris.

Cuvier, G. (1861). Éloge historique de Daubenton. *Recueil des éloges historiques, Firmin Didot Frères*, 1, 3–34.

Dambricourt Malassé, A. (1993). Continuity and discontinuity during modalities of hominization. In *Modes and Tempos in Evolution in the Quaternary*, Chaline, J., Werdelin L. (eds). *Quaternary International*, 19, 85–100.

Darwin, C. (1859). *On the Origin of Species by Means of Natural Selection, or the Preservation of Favoured Races in the Struggle for Life*. John Murray, London.

Darwin, C. (1868). *The Variation of Animals and Plants under Domestication*. John Murray, London.

Darwin, C. (1871). *The Descent of Man, and Selection in Relation to Sex*. John Murray, London.

Darwin, C. (1872). *The Descent of Man, and Selection in Relation to Sex*, 2nd edition. John Murray, London.

Darwin, C. (1873–1879). Correspondence with de Quatrefages [Online]. Available at: https://www.darwinproject.ac.uk/search?keyword=armand%20de%20Quatrefages;f1—addressee=Quatrefages%20de%20Br%C3%A9au,%20Armand%20de%20Quatrefages.

Darwin, C. (1875). Correspondence with Francis Galton [Online]. Available at: https://galton.org/letters/darwin/correspondence.htm.

Daubenton, L. (1764). Mémoire sur les différences de la situation du grand trou occipital dans l'homme et les animaux. *Histoire de l'Académie Royale des Sciences avec les Mémoires de Mathématique et de Physique pour la même année*, 568–575.

Daudin, H. (1926). *De Linné à Jussieu, Méthodes de la classification et idée de série en botanique et en zoologie (1740–1790)*. F. Alcan, Paris.

Delon, M. (n.d.). À propos de l'œuvre [Online]. Available at: https://gallica.bnf.fr/essentiels/buffon/histoire-generale-de-la-nature.

Desanti, J.-T. (2005). Réflexion sur le concept de Mathesis. *Figures de la psychanalyse*, 2(12), 103–137.

Diderot, D. and Le Rond d'Alembert, J-B. (1762). *Les sciences, les arts libéraux, et les arts méchaniques, avec leur explication ; Agriculture et Economie rustique*. Briasson, David, Le Breton, Durand, Paris.

Dupont, E. (1866). Etude sur les fouilles scientifiques exécutées pendant l'hiver de 1865–1866 dans les cavernes des bords de la Lesse. *Bulletins de l'Académie royale de Belgique*, 2(13), 7.

Élien, C. (2001–2002). *La personnalité des animaux, Livres I à IX, traduit et commenté par Arnaud Zucker*. Belles Lettres, Paris.

Fabrici d'Acquapendente, G. (1600). *De formato foetu*. Francesco Bolzetta, Venice.

Fick, L. (1853). Uber die Architektur des Schädels der Cerebrospinal Organismen. *Archiv für Anatomie, Physiologie und wissenschaftliche Medicin*, 86–136 [Online]. Available at: https://www.biodiversitylibrary.org/item/49684#page/676/mode/1up.

Fischer, J.-L. (2003). *Buffon et les théories de la génération au 18ème siècle. La pluridisciplinarité dans les enseignements scientifiques*, Volume 1. Histoire des sciences, Paris.

Foucault, D. (1999). Au déclin de l'averroïsme latin, Giulio Cesare Vanini (1585–1619). Horizons Maghrébins. *Le droit à la mémoire*, 40, 69–80.

Franzen, J.L. (2000). Auf dem Grunde des Urrheins. Ausgrabungen bei Eppelsheim. *Natur und Museum*, 130(6), 169–180.

Fréart, R. (1651). Traitté de la peinture. *Donné au public et traduit d'italien en françois*. Jacques Langlois, Paris.

French, R.K. (1985). Berengario da Carpi and the use of commentary in anatomical teaching. In *The Medical Renaissance of the Sixteenth Century*, Wear, A., Frenchand, R.K., Lonie, I.M. (eds). Cambridge University Press, Cambridge.

Galton, F. (1875). A theory of heredity. *Journal of the Anthropological Institute*, 5, 329–348.

Gaudant, J. (1992). Hommage à Richard Owen (1804–1892), le Cuvier anglais. *Travaux du Comité français d'histoire de la géologie*, 3(4), 109–117.

Gaudant, J. (2008). Albert Gaudry (1827–1908), paléontologue darwinien ? *Travaux du Comité français d'histoire de la géologie*, 3(22), 105–111.

Geoffroy Saint-Hilaire, E. (1795). Mémoire sur les rapports naturels des Makis Lémur. *Magasin Encyclopédique*, 1, 20–50.

Geoffroy Saint-Hilaire, E. (1818). Philosophie anatomique. *Pièces osseuses des organes respiratoire sous le rapport de la détermination et de l'identité de leurs pièces osseuses*, Volume 1. J.B. Baillière, Paris.

Geoffroy Saint-Hilaire, E. (1822). Philosophie anatomique. *Des monstruosités humaines*, Volume 2. J.B. Baillière, Paris.

Geoffroy Saint-Hilaire, I. (1832–1837). *L'Histoire générale et particulière des anomalies de l'organisation*. Librairie royale de l'Académie de Médecine. J.B. Baillière, Paris.

Geoffroy Saint-Hilaire, I. (1852). Sur le gorille. *Comptes Rendus de l'Académie des Sciences*, 34, 81–84.

Geoffroy Saint-Hilaire, E. and Cuvier, G. (1795a). Histoire naturelle des orangs-outangs. *Magasin encyclopédique*, 3, 451–463.

Geoffroy Saint-Hilaire, E. and Cuvier, G. (1795b). Mémoire sur une nouvelle division des mammifères, et les principes qui doivent servir de base dans cette sorte de travail. *Journal des Sciences, des Lettres et des Arts*, 2, 164–190.

Georges, H. (1892). Le crâne de Canstadt. *Bulletins et Mémoires de la Société d'Anthropologie de Paris*, 3, 365–376.

de Germiny, G. (1933). Ambroise Paré, zoologiste. Paris, Société nationale de protection de la nature et d'acclimatation de France, Paris.

Gilbert, S.F. (2004). *Biologie du développement*. Editions de Boeck Université, Brussels.

Gesner, C. (1545–1555). *Bibliotheca universalis sive catalogus omnium scriptorum locupletissimus in tribus linguis Latina, Graeca et Hebraica*. Froschauer, Zurich.

Giraud-Soulavie, J.-L. (1780–1784). *Histoire naturelle de la France méridionale*. Belin, Paris.

von Goethe, J.W. (1786). *Über den Zwischenkiefer des Menschen und der Thiere*. Mit 5 gefalt. Kupfertafeln von Kasimir van de Velde, Iena.

von Goethe, J.W. (1831). *Über den Zwischenkiefer des Menschen und der Thiere*. Verh. Verhandlungen der Kaiserlichen Leopoldinisch-Carolinischen Deutschen Akademie der Naturforscher, Halle.

Gohau, G. (1994). Analyse d'ouvrage G. W. Leibniz Protogaea – de l'aspect primitif de la terre. *Travaux du Comité français d'histoire de la géologie*, 3(8).

Goujet, C.P. (1758). *Mémoire historique et littéraire sur le Collège royal de France*. A.M. Lautin, Paris.

Gratiolet, L.P. (1854). *Mémoire sur les plis cérébraux de l'homme et des primates*. A. Bertrand, Paris.

Gozzoli, A. (n.d.). Scientific itinerary in Toscany, Ospedale di Santa Maria Nuova [Online]. Available at: https://brunelleschi.imss.fi.it/itineraries/place/OspedaleSantaMariaNuova.html.

Haeckel, E. (1866). *Generelle Morphologie der Organismen. Allgemeine Grundzüge der Organischen Formen-Wissenschaft, Mechanisch Begründet durch die von Charles Darwin Reformirte Descendenz-Theorie*. Druck und Verlag, Berlin [Online]. Available at: https://books.google.fr/books/content?id=4x80AQAAMAAJ&hl=fr&pg=PT3&img=1&zoom=3&sig=ACfU3U0OY5pMJH13tDFobTh-uoHy74M0hQ&ci=29%2C82%2C905%2C1321&edge=0.

Haeckel, E. (1874). *Histoire de la création des êtres organisés d'après les lois naturelles*. C. Reinwald et Cie, Paris.

Haeckel, E. (1875). *Natürliche Schöpfungsgeschichte*. Georg Reimer, Berlin.

Haeckel, E. (1877). *Anthropogénie ou histoire de l'évolution humaine. Leçons familières sur les principes de l'embryologie et de la phylogénie humaine*. C. Reinwald et Cie, Paris.

Haeckel, E. (1897). *Le monisme, lien entre la religion et la science, profession de foi d'un naturaliste*. Schleicher frères, Paris.

Haeckel, E. (1903). *Anthropogenie oder Entwickelungsgeschichte des Menschen. Keimes- und Stammes-Geschichte*, 5th edition. Engelmann, Leipzig.

Harvey, W. (1628). *Exercitatio Anatomica de Motu Cordis et Sanguinis in Animalibus*. Guilielmi Fitzeri, Francofurti.

Heinz, O. (1991). *Rudolf Virchow. Begründer der Zellularpathologie. Eine Dokumentation*. Echter Verlag, Wurtzbourg.

Huard, P. (1961). Paul Broca (1824–1880). *Revue d'histoire des sciences*, 14(1), 47–86.

Huxley, T.H. (1863). *Evidence as to Man's Place in Nature*. Williams and Norgate, London.

Jufresa, M. (1995). Savoir féminin et sectes pythagoriciennes. *Clio, histoire, femmes et sociétés*, 2, 17–40.

Kahn, D. (2007). Plantes et médecine, (al)chimie et libertinisme chez Guy de la Brosse [Online]. Available at: https://www.biusante.parisdescartes.fr/histoire/medica/presentations/brosse.php.

Kegan, P. (1901). *The Thoughts of Blaise Pascal, translated from the text of M. Auguste Molinier*. George Bell and Sons, London.

Kielmeyer, K. (1995). Über die Verhältnisse der organischen Kräfte unter einander in der Reihe der verschiedenen Organisationen, die Gesetze und Folgen dieser Verhältnisse (1793). *Zeitschrift für Allgemeine Wissenschaftstheorie*, 26(1), 191–198.

de La Roche, D. and Petit-Radel, P. (1792). Encyclopédie méthodique. *Médecine*, Volume 5. Panckoicke, Paris.

Lamarck, J.-B. (1794). *Recherches sur les causes des principaux faits physiques (...) enfin de l'entretien de la vie des êtres organiques, de leur accroissement, de leur vigueur, de leur déperissement et de leur mort*. Maradan, Paris.

Lamarck, J.-B. (1802). *Recherches sur l'organisation des corps vivants et particulièrement sur son origine, sur la cause de ses developpements et des progrès de sa composition*. Chez l'Auteur au Muséum d'Histoire Naturelle, Paris.

Lamarck, J.-B. (1809). *Philosophie zoologique ou exposition des considérations relatives à l'histoire naturelle des animaux*. Dentu, Paris.

Lamartine, A. (1830). *Harmonies poétiques et religieuses*. Charles Gosselin, Paris,

Lamy, E. (1930). *Les cabinets d'histoire naturelle en France au XVIIIe siècle et le Cabinet du roi (1635–1793)*. Chez l'auteur, Paris.

Leguebe, A. (1958). Edward Tyson (1651–1708) et les débuts de la primatologie comparée. *Bulletin de la Société Royale Belge d'Anthropologie et de Préhistoire*, 69, 139–150.

Leibniz, W.G. (1689–1690). Dynamica de potentia et legibus naturae corporae. Gottfried Wilhem Leibniz Bibliothek, Hanover.

de Libera, A., Brenet, J.-B., Rosier-Catach, I. (2019). *Dante et l'Averroïsme*. Collège de France, Paris.

Linnaeus, C. (1735). *Systema Naturae, sive Regna Tria Naturae Systematice Proposita per Classes, Ordines, Genera, and Species*. Theodor Haak, Leiden.

Lorenza, D. (2000). Léonard de Vinci. Artiste et scientifique. *Les génies de la science, Pour la science*, 3.

Lortet, L.C. (1905). Les momies animales de l'ancienne Égypte. *Revue des deux mondes*, 5(27).

Margolin, J.-C. (1998). Cardan Jérôme. In *Dictionnaire des philosophes*, Encyclopaedia Universalis (ed.). Albin Michel, Paris.

Mendel, G. (1866). Versuche über Pflanzenhybriden. *Verhandlungen des naturforschenden Vereines zu Brünn*, IV (1865), 3–47.

Milne-Edwards, H. (1867). *Rapport sur les progrès récents des sciences zoologiques en France*. Imprimerie impériale, Paris.

dei Mondino, L. (1551). *Mathaei curtii, Papiensis Medici celeberrimi in Mundini, Anatomen commentarius elegans & doctus*. Apud Paganum Theobeldum, Lyon.

Owen, R. (1848). *On the Archetype and Homologies of the Vertebrate Skeleton*. John Van Voorst, London.

Paracelsus (1589–1595). *Œuvres, Tome VIII*. Huser de Brisgau, Basel.

Paré, A. (1585). *Les oeuvres d'Ambroise Paré divisées en vingt huict livres avec les figures et portraicts, tant de l'anatomie que des instruments de chirurgie, et de plusieurs monstres, reveuës et augmentées par l'autheur*. Gabriel Buon avec privilege du Roy, Paris [Online]. Available at: https://gallica.bnf.fr/ark:/12148/bpt6k15199224/f11.planchecontact.

Parent, A. (2019). Berengario da Carpi and the Renaissance of Brain Anatomy. *Frontier in Neuroanatomie*, 13(11), eCollection 2019.

Pascal, B. (1897). *Pensées*. Léon Brunschvicg, Paris.

Pélandan, J. (1910). *Les Manuscrits de Léonard de Vinci, les 14 manuscrits de l'Institut de France*. Bibliothèque internationale d'édition, Paris.

Pietsch, T.-W. (2012–2015). *L'histoire des sciences naturelles de Cuvier, dix-neuf leçons sur les seizième et dix-septième siècles*. Muséum national d'Histoire naturelle, Paris.

Platter, F. (1583). *De corporis humani structura et usu, tabulis methodice explicati, iconibus a Cute illustrati*. Officina Frobeniana, Ambrosius Frobenius, Basel.

Pointe, J.P. (2020). Au chevet des Œuvres d'Ambroise Paré [Online]. Available at: https://bibulyon.hypotheses.org/13979.

Portal, A. (1770). Histoire de l'anatomie et de la chirurgie, Tome 4. In *L'origine et les progrès de ces sciences*. Didot le Jeune, Paris.

Postec, A. (2014). Un exemplaire singulier du De animalibus d'Albert le Grand et son illustration. *Reinardus*, 26, 1, 137–160 [Online]. Available at: http://mediatheque-numerique.inp.fr/Conferences/Le-Commentaire-d-Albert-le-Grand-sur-les-animaux-d-Aristote-XIVe-siecle-latin-16169-l-illustration-au-service-de-la-pedagogie.

de Quatrefages, A. and Hamy, E. (1874). La race de Cro-Magnon dans l'espace et dans le temps. *Bulletins et Mémoires de la Société d'Anthropologie de Paris*, 9, 260–266.

Ravaisson-Mollien, C. (1881). *Les manuscrits de Léonard de Vinci. Les manuscrits de la Bibliothèque de l'Institut*. A. Quantin, Paris.

Redi, F. (1668). *Esperienze intorno alla generazione degl'insetti*. All'insegna della Stella, Firenze.

Richter, J.P. (1883). *The Literary Works of Leonardo da Vinci*. Low, London.

Roger, J. (1988). La place de Buffon dans l'histoire des Sciences de la Terre. *Travaux du Comité français d'Histoire de la Géologie*, 3(2), 81–88.

Ruel-Kellermann, M. and Leyrat-Cornuejols, M.-R. (2015). Introduction à la transcription et à la traduction du Libellus de Dentibus d'Eustache [Online]. Available at: https://www.biusante.parisdescartes.fr/eustache/intro.pdf.

Sartori, E. (2006). *Histoire des femmes scientifiques de l'Antiquité au XXème siècle*. Plon, Paris.

Schaaffhausen, H. (1858). Zur Kenntniss der ältesten Rassenschädel. *Archiv Verdinbung Mehrereh Gelehrten*, 453–488.

Serres, E. (1832). *Théorie des formations et des déformations organiques, appliquées à l'anatomie de Ritta-Christina et de la duplicité monstrueuse*. J.B. Baillière, Paris.

Servicen, L. (1942). *Les Carnets de Léonard De Vinci*. Gallimard, Paris.

Smith, J.E.H. (2007). La génération spontanée et le problème de la reproduction des espèces avant et après Descartes. *Revues Philosophiques*, 34(2), 229–436.

Taquet, P. (2006). *Georges Cuviee, naissance d'un génie*. Odile Jacob, Paris.

Taubenschlag, R. (1948). The law of Greco-Roman Egypt in the light of the papyri. *Revue belge de Philologie et d'Histoire*, 26(1–2), 259–260.

Tulpius, N. (1641). *Amstelredamensis Observationes Medicae*. Elzevier, Amsterdam.

Tyson, E. (1699). *Orang-outang, sive Homo Sylvestris, or, the Anatomy of a Pygmie Compared with that of a Monkey, an Ape, and a Man*. Thomas Bennet and Daniel Brown, London.

Van Leeuwenhoek, A. (1669). *Historia Insectorum Generalis*. Menardus van Dreunen, Utrecht.

Velpeau, A. (1833). *Embryologie ou ovologie humaine, contenant l'histoire descriptive et iconographique de l'œuf humain*. Bailliere, Libraire de l'Académie royale de médecine, Paris [Online]. Available at: https://gallica.bnf.fr/ark:/12148/bpt6k96121668/f144.item.r=génésis.

Vemaas, K. (2017). Map of the Presocratic Greek world [Online]. Available at: https://commons.wikimedia.org/wiki/File:Map_of_the_Presocratic_Greek_World.png.

Vernant, J.-P. (2018). Les origines de la philosophie [Online]. Available at: https://blogs.mediapart.fr/segesta3756/blog/270918/les-origines-de-la-philosophie-par-jean-pierre-vernant-0.

Vesale, A. (1555). *De chirurgia scriptores optimi*. Oporinus, Bâle.

Vialleton, L. (1908). *Un problème de l'évolution ; la théorie de la récapitulation des formes ancestrales au cours du développement embryonnaire (loi biogénétique fondamentale de Haeckel)*. Coulet et fils, Montpellier.

Vialleton, L. (1929). *L'origine des êtres vivants, l'illusion transformiste*. Plon, Paris.

de Vinci, L. (1480–1518). Codex Arundel, MS 263, folio 236v. La planète magnifique, folio 233 recto [Online]. Available at: https://www.scirp.org/pdf/AHS_2019091616153569.pdf.

de Vinci, L. (1485). Codex Atlanticus, p. 878. [Online]. Available at http://codex-atlanticus.it/#/Detail?detail=663.

de Vinci, L. (1485–1490). Codex Windsor [Online]. Available at: https://www.rct.uk/collection/search#/2/collection/919057/recto-the-skull-sectioned-verso-the-cranium.

de Vinci, L. (1487–1505). Codex Forster [Online]. Available at: https://www.vam.ac.uk/articles/explore-leonardo-da-vinci-codex-forster-i#?c=&m=&s=&cv=1&xywh=-1%2C-40%2C3004%2C2069.

Vinci, L. (1487–1508). Manuscrits du Codex de la Bibliothèque de l'Institut de France [Online]. Available at: https://www.photo.rmn.fr/CS.aspx?VP3=SearchResult&VBID=2CO5PCXE2R30U&SMLS=1&RW=1536&RH=752.

de Vinci, L. (1490–1505). Deux Codex de Madrid à la bibliothèque nationale [Online]. Available at: http://leonardo.bne.es/index.html.

de Vinci, L. (1505). *Codex on the Flight of Birds*. Royal Library of Turin, Turin.

de Vinci, L. (1508–1510). *Codex Leicester*. Bill Gates Collection, Seattle.

de Vinci, L. (2010). Les carnets de Léonard. *Le journal du CNRS* [Online]. Available at: https://lejournal.cnrs.fr/infographies/dans-les-carnets-de-leonard

Virchow, R. (1857). *Untersuchungen über die Entwicklung des Schädelgrundes im gesunden und krankhaften Zustand und über den Einfluß derselben auf Schädelform.* Gesichtbildung und Gehirn Reimer, Berlin [Online]. Available at: https://books.google.cd/books?id=Qvh3hkmsHXEC&pg=PP5&hl=fr&source=gbs_selected_pages&cad=2#v=onepage&q&f=false.

Vons, J. (2016). Guy Patin, une exception parmi les médecins du Collège royal au XVIIème siècle ? *Histoire des sciences médicales*, 1(4).

Wagner, A. (1839). Fossile überreste von einem Affen und einigen andern Saugethieren aus Griechenland. *Munchener Gelehrte Anzeigen*, 8.

Weil, S. (1947). *La pesanteur et la grâce*. Plon, Paris.

Weismann, A. (1892). *Essais sur l'hérédité et la sélection naturelle*. C. Reinwald & Cie, Paris.

Welcker, H. (1862). *Untersuchungen über Wachstum und Bau des menschlichen Schädels.* Engelmann, Leipzig [Online]. Available at: https://books.google.fr/books?id=Bk9KNrjWZCoC&pg=PA167&hl=fr&source=gbs_selected_pages&cad=2#v=onepage&q&f=false.

de Wit, H. (1992). *Histoire du développement de la biologie*, volume 1. Presses polytechniques et universitaires romandes, Lausanne.

Wolff, C.F. (1768–1769). *De formatione intestinorum praecipue, tum et de amnio spurio, aliisque partibus embryonis gallinacei, nondum visis, observationes, in ovis incubatis institutae*. Academia Scientiarum, Petropoli [Online]. Available at: https://www.biodiversitylibrary.org/item/113647#page/645/mode/1up.

Index

H, I, J

L, M

N, O

P

Q, S